KEY NOTES ON
AGRONOMY

For Ready Reference to the

STUDENTS, TEACHERS, RESEARCHERS & ASPIRANTS OF COMPETITIVE EXAMINATIONS

THE EDITORS

Dr. U.D. Chavan obtained his M.Sc. (Agri. in Biochemistry) degree from Mahatma Phule Krishi Vidyapeeth, Rahuri. He received his Ph.D. degree in Food Science from Memorial University of Newfoundland St. John's Canada in 1999. He has done International Training on "Global Nutrition 2002" at Uppsala University Uppasala, Sweden in 2002. Dr. Chavan worked as Senior Research Assistant in the Department of Biochemistry & Food Science and Technology at MPKV Rahuri from 1988 to 2000. During his Ph.D., he worked as Technician/Research Associate at Atlantic Cool Climate Crop Research Center and Agriculture and Agri-Food Canada. He received D.Sc. degree in 2006 from USA.

Dr. Chavan is presently working as a Senior Cereal Food Technologist in the Department of Food Science & Technology at Mahatma Phule Krishi Vidyapeeth, Rahuri.

Dr. J.V. Patil obtained his M.Sc. (Agri.) from, MPKV, Rahuri. He completed his course work for Ph.D. at CCSHAU, Hisar and research at MPKV, Rahuri in 1992. He rendered his research and teaching services at MPKV Rahuri as Geneticist, Associate Professor, Plant Breeder and Professor of Genetics & Plant Breeding and Head, Genetics and Plant Breeding Department, MPKV, Rahuri. He also delivered many administrative responsibilities in the University. Dr. Patil joined as the Director, Directorate of Sorghum Research, Hyderabad in August 2010.

THE CONTRIBUTORS

Dr. V.T. Jadhav is an Assistant Professor in the Department of Agronomy at Mahatma Phule Krishi Vidyapeeth, Rahuri.

Dr. C.B. Gaikwad is Head, Department of Agronomy at Mahatma Phule Krishi Vidyapeeth, Rahuri.

Dr. H.M. Patil is an Associate Professor in the Department of Agronomy at Mahatma Phule Krishi Vidyapeeth, Rahuri.

KEY NOTES ON
AGRONOMY

For Ready Reference to the

STUDENTS, TEACHERS, RESEARCHERS & ASPIRANTS OF COMPETITIVE EXAMINATIONS

Editors

U.D. CHAVAN
&
J.V. PATIL

Contributors

V.T. JADHAV
C.B. GAIKWAD
H.M. PATIL

2015

Daya Publishing House®
A Division of
Astral International (P) Ltd
New Delhi 110 002

Published by	:	**Daya Publishing House®**
		A Division of
		Astral International Pvt. Ltd.
		– ISO 9001:2008 Certified Company –
		4760-61/23, Ansari Road, Darya Ganj
		New Delhi-110 002
		Ph. 011-43549197, 23278134
		E-mail: info@astralint.com
		Website: www.astralint.com
Laser Typesetting	:	**Twinkle Graphics, Delhi**
Printed at	:	**Thomson Press India Limited**

PRINTED IN INDIA

PREFACE

India is an agricultural country. The Indian economy is basically agarian. Inspite of economic and industrialization, agriculture is the backbone of the Indian economy. As Mahatma Gandhi said "India's lives in villages and agriculture is the soul of Indian economy". Agriculture is a vast subject and encompasses at least 20 major and minor subjects in it. New developments have lead to entirely a new face of agriculture. Study of agriculture has always been intrigued with a mosaic of interwove concepts, subjects, facts and figures. There are number of books and large literature on Agronomy but the Key Notes type of book have not been compiled in a readable manner.

The present book *"Key Notes on Agronomy"* has been designed to fulfill this long felt need of students, teachers, researchers and aspirants of competitive examinations. It is designed in such a way that give rapid, easy access to the core materials in a short format which facilitates easily learning and rapid revision. The book carries fundamentals of Agronomy. There are five chapters elaborating Discoveries, Abbreviations, Terminology, Differences/comparison as well as references also included. The most recent information is provided along with a detailed list of references for further reading.

Hope this book would be highly useful for graduate and post-graduate students of agriculture, teachers and researchers. This book will also useful for the aspirants of various competitive examinations such as Agricultural Research Service (ARS), ICAR- National Eligibility Test (NET), State Eligibility Test (SET), Junior Research Fellowship (JRF), Senior Research Fellowship (SRF), Civil Services, Allied Agricultural Examinations and Extension Workers for reference and easy answers of many complicated questions. Thus it is expected that this book will adequately meet the need of wider circle of students and readers for preparing their professional career.

We acknowledge the references that are used in this manuscript. Authors are also thankful to all scientists and friends who have helped directly or indirectly while preparing this manuscript. The editors of grateful to all the contributors for their cooperation, support and timely submission of their manuscripts for bringing out this publication. We would have like to acknowledge the patience and support of our families whilst we have spent many hours with drafts of manuscripts rather than with them. Lastly, our sincere thanks to publisher Astral International Pvt. Ltd., New Delhi who provides an opportunity to publish this book.

To all readers we extend an invitation to report that no doubts have escaped our attention and to offer suggestion for improvements that can be incorporated in future editions.

U.D.Chavan and J.V. Patil

Editors

CONTENTS

Preface   *(v)*

1. Discoveries   1

2. Abbreviations   2

3. Terminology   10

4. Differences/Comparison   279

5. Reasoning   282

References   306

1

DISCOVERIES

Essential nutrients, forms of uptake and discovery.

Year of discovery	Scientist	Nutrient and symbol	Principal form for uptake
1804	T De Saussure	Oxygen O	H_2O, O_2
1856	F Salm-Horstmar	Calcium Ca	Ca^{2+}
1860	J Sachs	Iron Fe	Fe^{2+}, Fe^{3+}
1872	GK Rutherford	Nitrogen N	NH_4^+, NO_3
1882	J Sachs	Hydrogen H	H_2O
1882	J Sachs	Carbon C	CO_2
1890	AFZ Schimper	Potassium K	$K+$
1903	Postemak	Phosphorus P	$H_2PO_4^-$, HPO_4^{2-}
1906	Willstatter	Magnesium Mg	Mg^{2+}
1911	Peterson	Sulphur S	SO_4^{2-}, SO_2
1922	JS Mc Hargue	Manganese Mn	Mn^{2+}
1923	K Warington	Boron B	H_3BO_3
1926	AL Sommer and CB Lipman	Zinc Zn	Zn^{2+}
1931	CB Lipman and G Mac Kinney	Copper Cu	Cu^{2+}
1938	DI Amon and PR Stout	Molybdenum Mo	MoO_4^{2-}
1954	TC Broyer *et al*	Chlorine Cl	Cl^-

2

ABBREVIATIONS

Abbreviation	Full form
IUCN	International Union for Conservation of Nature and Natural Resources
ABA	Abscisic Acid
ADF	Acid Detergent Fiber
ADP	Adenosine Diphosphate
AGR	Absolute Growth Rate
AI	Activity Index
AICMIP	All India Coordinated Maize Improvement Project
AICRIP	All India Coordinated Rice improvement project
AICRPDA	All India Coordinated research Project for Dryland Agriculture
AICSIP	All India Coordinated Sorghum Improvement Project
ANOVA	Analysis of Variance
ASM	Available Soil Moisture
ASN	Ammonium Sulphate Nitrate
ATER	Area-Time Equivalent Ratio
ATP	Adenosine Triphosphate
AUM	Animal Unit Month
BBF	Broad Bed and Furrow
BD	Bulk Density
BGA	Blue Green Algae
BOAA	Beta-N-oxalyl-Amino Alanine
BSDM	Brown Strip Downy Mildew
CAC	Cation Exchange Capacity
CAM	Crassulacean Acid Metabolism

Abbreviation	Full form
CAN	Calcium Ammonium Nitrate
CCC	Cycocel
CCS	Conventional Cropping System
CDA	Controlled Droplet Application
CEC	Cation Exchange Capacity
CEY	Crop Equivalent Yield
CF	Crude Fiber
CFC	Chlorofluoro Carbons
CGR	Crop-Growth Rate
CI	Competition Index
CII	Cropping Intensity Index
CIMMYT	International Maize and Wheat Center
CLUI	Cultivated Land Utilization Index
Cm	Centimeter
CMS	Cytoplasmic Male Sterility
CP	Crude Protein
CPE	Cumulative Pan Evaporation
CPRI	Central Potato Research Institute
CRI	Crown Root Initiation Stage
CRIDA	Central Research Institute for Dryland Agriculture
CRIJAF	Research Institute for Jute and Allied Fibers
CRRI	Central Rice Research Institute
CU	Consumption Use
CV	Coefficient of Variation
DAE	Days After Emergence
DAP	Diammonium Phosphate
DAS	Days After Sowing
DCA	Direct Contact Application
DCP	Digestible Crude Protein
DI	Diversity Index

Abbreviation	Full form
DNA	Deoxyribose Nucleic Acid
DPD	Diffusion Pressure Deficit
DW	Dry Weight
EC	Electrical Conductivity
EC	Emulsifiable Concentrate
ECP	Exchangeable Cation Percentage
EE	Ether Extract
EIL	Economic Injury Level
EMR	Electromagnetic Radiation
Ep	Elasticity of Production
Ep	Project Efficiency
ER	Effective Rainfall
ESP	Exchangeable Sodium Percentage
ET	Evapo-Transpiration
ETL	Economic Threshold Level
EW	East-West
FC	Field Capacity
FCV	Flu Cured Virginia
FI	Floral Initial
FIRB	Furrow Irrigated Raised Bed
FP	Farmers Practice
FSA	Farming System Analysis
FSAR	Farming System Adaptive Research
FSBDA	Farming System Baseline Data Analysis
FSCR	Farming System Component Research
FSP	Farming System Perspective
FSR	Farming System Research
FSRAD	Farming System Research and Agricultural Development
FV	Feeding Value
FW	Fresh Weight

Abbreviation	Full form
FYM	Farm Yard Manure
GOI	Government of India
GSD	Grassy Shoot Disease
HDI	Harvest Diversity Index
HI	Harvest Index
HPS	High Protein Selects
HWT	Hot Water Treatment
HYVs	High Yielding Varieties
IAA	Indole Acetic Acid
IARI	Indian Agricultural Research Institute
IBDU	Isobutylidene Diurea
ICARDA	International Center for Agriculture Research in Dry Areas
ICBR	Incremental Cost Benefit Ratio
IDM	Integrated Disease Management
IDRC	International Development Research Council
IER	Income Equivalent Ratio
IFS	Integrated Farming System
IMD	Indian Meteorological Department
INM	Integrated Nutrient Management
INSAT	Indian National Satellite System
IPM	Integrated Pest Management
IPNSS	Integrated Plant Nutrient Supply System
IPNS	Integrated Plant Nutrition System
IR	Irrigation Requirement
IRRI	International Rice Research Institute
IV	Importance Value
IW	Irrigation Water
IWM	Integrated Weed Management
IWMS	Integrated Weed Management System
k	Conductivity Cell-Constant

Abbreviation	*Full form*
KLS	Karnataka Light Soils
kp	Pan Coefficient
KVKs	Krishi Vigyan Kendras
LACE	Large Area Cropping Experiments
LAI	Leaf-Area Index
LAR	Leaf-Area Ratio
LER	Land Equivalent Ratio
LI	Lint Index
LLP	Lab-to-Land Programme
MAI	Monetary Advantage Index
MAI	Moisture Available Index
MCC	Maximum Capillary Capacity
MCI	Multiple Cropping Index
MDI	Moisture-Deficit Index
ME	Metabolizable Energy
MEY	Maximum Economic Yield
MHAT	Moist Hot Air Treatment
MOP	Muriate Of Potash
MP	Marginal Product
MRL	Maximum Residue Limit
MSS	Multi Spectral Scanner
MT	Million Tonnes
NAR	Net Assimilation Rate
NARP	National Agricultural Research Project
NBU	Neem Blended Urea
NCU	Neem Coated Urea
NCCU	Neem Cake-Coated Urea
NDF	Neutral Detergent Fiber
NE	Net Energy
NFE	Nitrogen-Free Extract

Abbreviation	Full form
NFSD	New Farming System Development
NIR	Net Irrigation Requirement
NLS	Northern Light Soils
NR	Nitrate Reductase
NSC	National Seed Corporation
NUE	Nitrogen Use Efficiency
NVI	Nutritive Value Index
ODR	Oxygen Diffusion Rate
OP	Osmotic Pressure
ORP	Operational Research Project
PAR	Photo-Synthetically Active Radiation
PCNB	Penta-Chlorine Nitrobenzene
PD	Particle Density
PDS	Public Distribution System
PE	Production Efficiency
PER	Protein Efficiency Ratio
PET	Potential Evapo-Transpiration
PI	Pearling Index
PMC	Press Mud Cake
ppm	Part Per Million
PQ	Photosynthetic Quotient
PRPF	Plant Row Plough Furrow
PSB	Phosphate Solubilizing Bacteria
PSM	Potassium Solubilising Micro-organisms
PSP	Photoperiod Sensitive Phase
PUFA	Polyunsaturated Fatty Acids
PWP	Permanent Wilting Point
RBD	Randomized Block Design
RCII	Relative Cropping Intensity Index
RD	Relative Density

Abbreviation	Full form
RFLP	Restriction Fragment Length Polymorphism
RGR	Relative Growth Rate
RHC	Red Hairy Caterpillar
RLD	Root Length Density
RMO	Rapeseed Mustard Oil
RNA	Ribose Nucleic Acid
RQ	Respiratory Quotient
RSD	Ratoon Stunting Disease
RT	Relative Turgidity
RTD	Relative Temperature Disparity
RUI	Resource-Use Indices
RWC	Relative Water Content
RYT	Relative Yield Total
SAR	Sodium Adsorption Ratio
SAT	Semi-Arid Tropics
SAUs	State Agricultural Universities
SC	Solution Concentrate
SCI	Simultaneous Cropping Index
SCII	Specific Crop Intensity Index
SCU	Sulphur Coated Urea
SI	Simpson's Index
SIW	Suitability of an Irrigation Water
SLER	Staple Land Equivalent Ratio
SLS	Southern Light Soils
SOM	Soil Organic Matter
SOP	Sulphate of Potash
SP	Soluble Powder
SPAC	Soil-plant Atmosphere Continuum
SRI	System of Rice Intensification
SSC	State Seed Corporation

Abbreviation	Full form
SSP	Soluble Sodium Percentage
SSP	Single Superphosphate
STP	Space Transplanting
TBS	Traditional Black Soils
TDC	Tarai Development Corporation
TDN	Total Digestible Nutrients
TDR	Time-Domine Reflectometer
TGMS	Thermo-Sensitive Genetic Male Sterility
TPS	True Potato Seed
TW	Turgid Weight
USGS	Urea Super Granules
USLE	Universal Soil Loss Equation
USWB	United States Weather Bureau
VAM	Vesicular *Arbuscular Mycorrhiza*
WA	Wild Abortive
WCE	Weed Control Efficiency
WDI	Weed Dry matter in the Inter crop
WDS	Weed Dry matter in Sole crop
WEC	Wind-Energy Conversion
WI	Weed Index
WP	Wettable Powder
WR	Water Requirement
WSE	Weed Smothering Efficiency
WUE	Water Use Efficiency
WX	Waxy gene
ZREAC	Zonal Research and Extension Advisory Committee

3

TERMINOLOGY

Term	Terminology
Abatement	Refers to minimise pollution through reuse or waste treatment.
Abiotic factors	Physical, chemical and other non-living components of the environment such as light, temperature, moisture, nutrients and other edaphic factors.
Aboriculture	Cultivation of tree species for fruits, gums, mats etc.
Abscisic acid (ABA)	An important growth-inhibitor of plants, initially called dormin, has wide range of physiological effects including promotion of senescence and abscission, causing dormancy in buds and seeds, closure of stomata, retardation and inhibition of growth known as 'anti-gibberellins' since it inhibits gibberellins-stimulated growth.
Abscissa, (X-axis)	Horizontal axis of a graph.
Abscission	Detachment of fruit, leaf or other parts from a plant.
	The shedding of fruits leaves or stems from the parent plant.
Abscission layer	A zone of cells in the petiole or other plant structure whose cells separate and thereby bring about leaf fall, fruit drop etc.
Absolute growth rate (AGR)	The rate of increase in size of a growing plant (or part of it) in a given time under specific condition.
Absolute humidity	It is the amount of water vapour present in a unit volume of air.
Absolute transpiration	The rate of water loss from a plant determined experimentally.

Term	Terminology
Absolute water requirement	Also called consumptive use of water. This is the quantity of water in ha-cm per crop season absorbed by the crop together with the evaporation from the crop producing land. It includes the water used by evapotranspiration and retained in plant body.
Absolute weed	These are the plants which are undesirable regardless of time and place.
Absorbent	A substance which takes up liquids or gases and distributes them throughout its mass, *e.g.*, charcoal.
Absorption	The process by which a substance is taken into or includes within another substance, like intake of water by soil or intake of nutrients by plants.
	The process of penetration into the plant tissue by roots or foliage.
Absorption spectrum	The curve showing the per cent absorption of light at each wave length by a pigment.
Absorptivity coefficient	This is a decimal fraction expressing the portion of impinging radiation that is absorbed. A leaf, for example, has an absorptivity coefficient of about 0.98 in the far infrared portion of the spectrum.
Abundance (nuclear)	Relative concentration of the individual isotope in a mixture of isotopes (belonging to the same chemical element).
Abundance ratio	The proportion of various isotopes in an atom. The values are usually expressed as atomic percentage.
Acaricide	Any chemical agent that kills members of the Order Acarina (mites and ticks).
Accelerated erosion	When the vegetation is removed and the land put under cultivation the natural balance existing between the soil, its vegetational cover and climate is disturbed. The removal of the surface soil takes place at a much faster rate than it can cover is built up by the soil forming processes.
Accelerator (nuclear)	Commonly referred to as an 'atom smasher'. A device used to impart a high kinetic energy to a charged particle to cause it to undergo nuclear or particle reactions.

Term	Terminology
Acceptability	Readiness with which, animals select and eat a forage; sometimes used interchangeably to mean either palatability or voluntary intake.
Acclimatization	An adjustment in the morphology, or physiology of an organism in response to a change in the environment.
Accumulation	The uptake of substances against a concentration gradient and hence an energy-requiring process.
Acedophiles	Plant which can grow well under acidic soil conditions (especially weeds).
Acetyl value	It is the amount of potassium hydroxide required in milligram to neutralize the acetic acid liberated by the, hydrolysis of one g of the acetylated substance.
Achene	A dry indehiscent, one-seeded fruit in which the ovary wall remains free from the seed-coat.
Acid	A substance, which forms hydrogen ions (H^+) when dissolved in water; a proton donor.
Acid detergent fiber (ADF)	Lignin + cellulose.
Acid equivalent	The theoretical yield of parent acid from an active ingredient in acid-based herbicides.
	The theoretical yield of parent acid from an active ingredient.
Acid rain	Rain water containing excessive concentration of acidic compounds, primarily NO_3^-, SO_4 and H^+, having pH of 5.0-5.5. Received where atmospheric pollution through industrial activity/vehicular exhaust is high.
Acid salts	When only part of the replaceable hydrogen in an acid is replaced by a base an acid salt is obtained. The acid salt possesses the properties of salt in addition to those of acids. They react with alkalies to give normal salt and water (acid salt, $NaHSO_4$, $NaHCO_3$ NaH_2PO_4).
Acid soil	A soil with a pH reaction of less than 7.0 (usually less than 6.6). An acid soil has a preponderance of hydrogen ions over hydroxyl ions, and blue litmus paper turns 'red in contact with moist acid soil.

Term	Terminology
Acid-forming fertilizer	A fertilizer which leaves acid residue in the soil applied.
Acidity	The state of being acid in reaction, *i.e.,* pH less than 7.0 which would be evident to the taste as sourness.
Acidity of a base	The number of hydrogen atoms with which one molecule of a base reacts when it combines, *e.g.,* NaOH is monoacid base; $Ca(OH)_2$ is a diacid base and $Cr(PH)_3$ is a triacid base.
Acidity potential	The amount of exchangeable hydrogen ion in a soil that can be rendered free or active in the soil solution by cation exchange. Usually expressed in milliequivalents per unit mass of soil.
Acidophilus	The weeds which grow well characteristically in acidic soils having pH between 4.5 and 6.5.
	The plants which can grow in acidic soils (pH range 4.5-6.5).
Acquired character	A modification or variation in a character arising from the influence of the environmental factors during the course of development of the organism and not by hereditary factors.
Acquired immunity	Specific immunity acquired through exposure to the antigen.
Acre	It is a unit of measurement of land area, equivalent to 4,840 sq. yards or 0.4047 of a hectare.
Acre foot of water	It is the amount of water that would cover an acre of land to a depth of one foot, assuming no seepage, evaporation arid run-off losses.
Acre-inch	It is a measure of quantity of flow of water and is equal to a flow which will cover 1 acre to a depth of 1 inch.
Acropetal	Towards the apex of a plant organ; generally upwards in shoots and downwards in roots. Opposite of basipetal.
Actinium series	A natural radioactive series.

Term	*Terminology*
Actinomycetes	A general term to cover a group of organisms intermediate between the bacteria and the true fungi, usually producing a branched mycelium and which sporulate by segmentation of the entire mycelium or more commonly by segmentation of special hyphae. Any organism belonging to the order Actinomycetales.
Activated sludge	Sludge that has been aerated and subjected to bacterial action.
Activation (nuclear)	The process of causing a substance to become artificially radioactive by subjecting it to bombardment by neutrons or other particles.
Activation analysis	An analytical procedure permitting the detection and measurement of trace quantities of elements following their exposure to flux of neutrons.
Activator	A substance that accelerates the effect or increases the total effect of a product.
Active absorption	Absorption of water and other substances against the concentration gradient and therefore, involving expenditure of energy in contrast to imbibitions and osmosis.
Active acidity	The activity of hydrogen ions in the aqueous phase of soil. Its activity is measured and expressed as pH value.
Active ingredient	The chemical in a product that is principally responsible for herbicidal effects.
	The active component of a formulated product such as fungicide, *e.g.,* Ridomil 25 wp and 75% inert material. The Ridomil has 25% a.i.
Activity (nuclear)	The strength of a radioactive source. In absolute unit, it refers to the numbers of radioactive atoms decaying per unit of time. In relative terms it is expressed in terms of the number of recorded events per unit of time. Also a synonym for radioactivity. Absolute activity is usually expressed in curies or millicuries.

Term	Terminology
Activity index (AI)	Active index is used by the Association of Official Analytical Chemists (AOAC) to evaluate the solubility of urea formaldehyde compounds.
Actual crop evapo-transpiration	Rate of evapo-transpiration equal to or smaller than potential transpiration evapo-transpiration as affected by the level of available soil water, salinity, field size, or other causes; mm/day.
Actual farm yield	The yield of a crop or variety on the farmer's field under optimum management practices.
Actual vapour pressure	Pressure exerted by water vapour contained in the air; millibar (mb) or mm of HG.
Ad libitum feeding	Where animals are permitted to eat daily as much as they desire.
Adaptation	An adjustment by an organism to environmental conditions.
Adaptation value	The survival and reproductive value of a genotype relative to others under particular enviornrnent.
Add sulphate soil	Very acid soil (pH < 4) in which sulphuric acid is formed by the oxidation of S-bearing, ferrous pyrites minerals. Found primarily in coastal, deltaic and estuarine areas of the humid tropics. Sometimes also called cat clays.
Add value	It is the amount of potassium hydroxide required in milligram to neutralize the free acid present in one g of the substance.
Additive	Substance intended to improve the properties of a fertilizer or soil conditioner.
Additive series	In intercropping, introduction of another plant species without reducing the population of the first species from the optimum.
Adenosine diphosphate (ADP)	A compound made up of adenine, 5 carbon sugar ribose and 2 phosphate groups. It involves in the mobilization of energy in cellular metabolism.
Adenosine triphosphate (ATP)	ADP with an additional phosphate group attached by a high-energy bond, its decomposition into ADP and phosphate results in the release of energy needed for cellular metabolism.

Term	Terminology
Adhesion	The force of attraction that binds the molecules of different kinds.
Adhesive	A material included in a formulation to increase its sticking power or adhesiveness. Also known as a sticker.
Adiabatic lapse rate	The rate at which temperature changes as air rises or falls is called adiabatic lapse rate. The rate is constant for dry air.
Adjacent area	Areas situated to the cropped area. This includes soil, snow or intercepted precipitation on the area in the particular locality which is considered for evapo-transpiration in computing the consumptive use of crop.
Adjuvant	An ingredient which when added to a formulation aids the action of the toxicant. Includes such materials as wetting agents, spreaders, adhesives, emulsifiers, dispersing agents and correctives.
	An additive ingredient that facilitates or modifies the action of the principle ingredient.
ADP (adinosinedi phosphate)	Adenosine-derived ester formed in plant cells and converted to ATP for energy storage.
Adsali sugarcane	Sugarcane which takes 18 months for harvesting, usually planted in June-July.
Adsorption	The attraction of ions or compounds to the surface of solid. Solid colloids adsorb large amounts of ions and water.
	The chemical or physical attraction, adhesion and accumulation of molecules at the soil-water or soil interface, resulting in one or more ionic or molecular layers on the surface of soil particles. It can refer to gases, dissolved substances or liquids on surface of solids or liquids.
Advance curve	The curve which shows relationship between distances travelled by a flow of water and time in surface irrigation methods.
Advance phase (irrigation)	It is the portion of the total irrigation time during which water advances in over land flow surface from the upper end of the field to the lower end of the field. The time elapsed during this phase is called advance time.

Term	Terminology
Advection	The process of transport of an atmospheric property (such as heat, water vapour, or momentum) solely by the horizontal motions of the atmosphere.
Advective energy	The energy developed from horizontal heterogeneity in climatic parameters or the energy brought in an area through horizontal movement of winds. (It is an important source of energy if winds are coming from hot and dry adjacent areas, especially in tropical hot summers which results in greater evapo-transpiration than normally due to solar energy).
Adventitious	Organs arising in unusual positions, as buds from roots.
Aeolin	A sediment deposit carried by wind.
Aeration (soil)	The process by which air and other gases in the soil are renewed. The rate of soil aeration depends largely on the size and number of soil pores and on the amount of water dogging to permit rapid aeration, is said to be well aerated, while a poorly aerated soil either has few large pores or has most of its pores blocked by water.
Aeration (storage)	The process of movement of air through grain in storage at low air flow rate for the primary purpose of maintaining or improving the quality of the grain.
Aeration porosity	The proportion of the bulk volume of the soil that is filled with air when the moisture tension is at some specified value. The tension is usually specified in the range 40 to 100 cm of water.
Aerenchyma	Aerating tissue in aquatic plant organs, which is characterized by large intercellular spaces.
Aerial	Living or occurring in air.
Aerial root	A root usually arising adventitiously from a stem. Epiphytes produce aerial roots which have specialized tissue called 'velaman' for absorbing moisture from the air. Aerial roots may contain chlorophyll and photosynthesize.
Aerial spraying	Application of pesticide or fertilizer in the form of spray by using aeroplane or helicopter, with the objective of covering vast area in a short time.

Term	Terminology
Aerobe	Any organism which grows in the presence of oxygen or air. An obligate aerobe grows only under such conditions.
Aerobic	(i) Conditions with oxygen gas as a part of the environment, (ii) living or active only in the presence of air or free oxygen, (iii) pertaining to the activity of organisms that grow under aerobic conditions, such as aerobic decomposition.
Aerobic respiration	Enzymatic oxidation of organic substances into simpler compounds like carbon dioxide and water with release of energy in the presence of oxygen.
Aerodynamic	It refers to forces of moving air acting upon the soil or crop surface.
Aerodynamically rough	A surface whose roughness elements are sufficiently large that surface the turbulent boundary layer reaches the surface.
Aerodynamically smooth	A surface whose roughness elements are sufficiently small to surface be entirely embedded in the laminar sub layer.
Afforestation	Artificial establishment of a forest on land which had not previously grown tree crops or on deforested land.
After ripening	A term for the collective physiological changes that occur in a dormant seed which makes it capable of germination.
After-cultivation	Harrowing, rolling, tilling and other cultivations carried out in a field after the crop has emerged.
Aftermath	The second growth of meadow plants after the first crop has been cut.
Aggregate (soil)	Many fine soil particles held in a single mass or cluster, such as a clod crumb, block or prism.
Aggregate sample	A combination of all increments from the sampling unit.
Aggressivity	It gives a simple measure of how much the relative yield increase in species 'a' is greater than that for species 'b' in an intercropping system and can be expressed as Aab.

Term	Terminology
Agricultural drought	It occurs where soil moisture and rainfall are inadequate during the growing season to support healthy crop growth to maturity and causes extreme crop stress and wilt.
Agricultural ecosystem or agro-ecosystem	It is composed of the total complex of the crops or animals in an area together with overall environment and as modified by management practices.
Agricultural forest	These are mixtures of tree crops, meadows, domestic animals, living fences (hedge and wind breaks), other perennial plants that produce food, fiber and chemicals. Thus, forests are designed to simulate the basic processes of a natural forest ecosystem for yielding human foods and materials.
Agricultural liming material	Material containing oxides, hydroxides and/or carbonates of calcium and/or magnesium, used for neutralizing the acidity of the soil.
Agricultural statistics	It concerns with scientific methods for collecting, organizing, summarizing, presenting and analyzing data generated from agricultural experiments as well as drawing valid conclusions and making reasonable decision on the basis of such analysis.
Agriculture	An activity of man, primarily aimed at the production of food, fiber, fuel etc. by optimum use of terrestrial resources.
Agri-silviculture	It is the conscious and deliberate use of land for the concurrent production of agricultural and forest crops.
	Growing of trees for timber, but with cultivated crops beneath
Agrobiology	A phase of the study of agronomy dealing with the relation of yield to the quantity of an added or available fertilizer element.
Agroclimatic regions	The grouping of different physical areas within the country into broadly homogeneous zones based on climatic and edaphic factors.
Agroclimatology	Study of those aspects of climate which are relevant to the problems of agriculture.

Term	Terminology
Agroecological zone	A major area of land that is broadly homogeneous in climatic and edaphic factors, but not necessarily contiguous, where a specific crop exhibits roughly the same biological expression.
Agroecology	The study of the relationship of agricultural crops and environment.
Agro-energy	It is a broad term used for referring to energy issues related to agriculture, both as a producer of energy as well as a user of energy. Sun's energy is captured and stored as biomass through the process of photosynthesis by plants which could be used as a direct source of energy in the form of food, feed and fuel or could be converted into gases or liquid forms of fuels for efficient use.
Agro-forestry	It is a self-sustaining land-management system which combines procluction of agricultural crops with that of tree crops as also with that of the livestock simultaneously or sequentially, on the same unit of land.
Agrology	The study of applied phases of soil science/and soil management.
Agromet station	It is an agricultural meteorological station which provides either meteorological and biological data or otherwise supplies data in establishing relation between weather and life of plants and animals.
Agrometeorology	A science dealing with climatologically conditions which have direct relation or relevance to agriculture.
Agronomic efficiency	Units of crop produced per unit input (nutrient) added.
Agronomy	The word has been derived from the Greek word 'agros' meaning fields and 'nomos' meaning management. It is a specialized branch of agriculture dealing with crop production and soil management or this is the application of scientific principles to the art of crop production.
Agro-pesticide	A term used to cover chemicals classified as pesticides, adjutants, curative agents and other chemicals, used to improve agricultural production by protecting crops or controlling pests.

Term	Terminology
Agro-silvipastoral system	A system in which land is managed for the concurrent production of agricultural and forest crops and for the rearing of domesticated animals. This system is, in fact a combination of agri-silviculture and the silvi-pastoral system.
Agrostology	A branch of science which deals with the study of grasses, their classification, management and utilization.
A-horizon	The surface horizon of a mineral soil having maximum biological activity or eluviations (removal of materials dissolved or suspended in water) or both.
Air capacity of soil	It is the quantity of air present in the soil at field capacity.
Air current	Vertical movement of air is known as air currents. Air-dry: State of dryness after prolonged exposure to air or any exposure sufficient to bring a material into moisture equilibrium with the air.
Air porosity	The proportion of the bulk volume of soil that is filled with air at any given time or under a given condition, such as a specified moisture tension.
Akiochi soils	These are flooded soils where hydrogen sulphide is formed due to sulphate reduction and anaerobic decomposition of organic matter.
Albedo	It is the ratio of the amount of visible light reflected by a body to the amount incident upon it. It is also called reflected radiation 'and expressed in percentage. It is measured with Albedometer.
Albuminous seed	A seed that contains endosperm at the time of germination, which nourishes the growing seedling.
Alfisols	Soils with grey to brown surface horizons, medium to high supply of bases, and B horizons of illuvial clay accumulation. These soils form mostly under forest or savanna vegetation in climates with slight to pronounced seasonal moisture deficit.
Algae	A heterogeneous group of photosynthetic unicellular and multicellular organisms comes under Thalophyta.

Term	Terminology
Alien weeds	When a weed is allowed to move from the place of its origin to a new area, and it establishes itself there, it becomes an introduced weed in its new environment. Such weeds are known as alien weeds or anthrophytes.
Aliphatics	The compounds derived from straight chain hydrocarbons.
Alkali soil	A soil with a high degree of alkalinity (pH 8.5 or higher) and with a high exchangeable sodium content (15% or more) with electrical conductivity less than 4 mmhos/cm at 25°C.
Alkaline (or basic) fertilizer	A fertilizer capable of increasing the residual alkalinity or reducing the residual acidity of the soil.
Alkaline soil	A soil with a pH above 7.0, usually above 7.3. Alkalization: The accumulation of sodium ions on the exchange sites in a soil.
Alkaloids	Bitter alkaline organic compounds containing carbon, hydrogen, nitrogen and in some oxygen. Familiar alkaloids are caffeine, morphine, nicotine, quinine and strychnine.
Allelochemicals	Certain crop residues produce allelochemicals tiring decomposition, if sufficient time is allowed to lapse between the harvests and sowing of the subsequent crop, the allelochemicals produced disappear.
Allelo-inhibition	Allelo chemicals inhibit more the growth of plants of species other than producer species.
Allelopathy	It is any direct or indirect harmful effect that one plant has on another or mutually on each other through the production of chemical compounds that escape into the environment.
	The phenomenon of one plant having detrimental effect on another through the production and release of toxic chemicals.
Alley cropping	A farming system in which arable crops are grown in alleys formed by trees or shrubs, established mainly to hasten soil-fertility restoration and enhance soil productivity.

Term	Terminology
Allophane	An alumino silicate mineral that has an amorphous or poorly crystalline structure and is commonly found in soils developed from volcanic ash.
Allowable soil-water depletion	Depth of soil water in the root zone readily available to the crop for given soil and climate allowing unrestricted evapo-transpiration as the fraction of total available soil water between field capacity and wilting point. Expressed as mm/m.
Alluvial soil	A soil developed from recently deposited alluvium and exhibiting essentially no horizon development or modification of the recently deposited materials.
Alluviation	The process of accumulation of gravel deposits, sand, silt or clay, at places in rivers, lakes, or estuaries where the flow is checked.
Alluvium	A general term for all detrital materials deposited or in transit by streams, including gravel, sand, silt day and all variations and mixtures of these.
Alpha decay	The radioactive transmutation of an element into another by the emission of an alpha particle.
Alpha particle	A particle which is identical to the helium nucleus, consisting of 2 protons and 2 neutrons. It carries a positive charge of 2.
Alpha radiation	The radiation consisting of alpha particles emitted by certain radioactive elements.
Alpha-ray	A ray of positively charged particles emitted during certain radioactive transformations.
Alternate drying and wetting (rice)	Frequent draining and re-flooding (aerobic and anaerobic cycle) in rice fields.
Alternate grazing	Grazing of two or more pastures. in succession.
Altitude	Vertical distance of a level, a point or an object considered, from a point measured from mean sea-level.
Aluminium toxicity	It is caused by excess water-soluble and exchangeable aluminium, usually on acid sulphate soils and strongly acidic soils. Symptoms are white or yellow interveinal blotches on the leaves. The leaves dry out and die. Roots are short and scanty and plants are stunted.

Term	Terminology
Alumino-silicates	Compounds containing aluminium, silicon and oxygen atoms as main constituents.
Aman rice	The word *'aman'* is derived from Arabic, meaning safety which indicates stability of the crop. *Aman* rice is direct seeded in pre-monsoon period (May-June) in the deep water (1-6 m deep) areas. The rice plants grow with rising flood water and settles at the ground where water subsides during the post-monsoon period. This rice crop is photo-sensitive and harvested during October-December.
Ambient temperature	The temperature of the surrounding atmospheric air.
Ambimobile	The character of chemical substances (*e.g.,* herbicides) which, when absorbed by the plant, move in both the phloem and xylem.
Ameliorants	Substances added to soil for the improvement of physical and chemical properties and to help in increasing crop yields.
Amelioration (land)	The process of enhancing the agricultural value of a land by drainage, tillage, liming, manuring, terracing etc.
Amendment	Any material such as lime, gypsum, sawdust, sulphur or soil conditioner that is added to the soil to improve the physico-chemical properties of the soil.
Amensalism	Symbiosis in which one organism is inhibited by the other but the latter remains unaffected.
Amensalistic polyculture	The interaction between crop species has a negative net effect on one species and no observable effect (negative or positive) on other species.
Amines	A class of compounds derived from ammonia by replacing the hydrogens with organic radicals.
Amino acids	These are building blocks of proteins and each amino-acid molecule contains amino group (~) and at least 1 carboxyl (–COOH) molecule.
Ammonia (liquid or anhydrous)	A material mostly produced by the synthetic process and obtained in the liquid form. It is used in the fertilizer industry for ammoniation of

Term	Terminology
	superphosphate, in making base mixtures, in making mixed fertilizer, or for conversion in to salts such as sulphate, phosphate, chloride, nitrate etc. It is also applied directly in to the soil by suitable mechanical means. Fertilizer grade anhydrous ammonia contains about 82 per cent of nitrogen.
Ammonia volatilization (soil)	Ammonia state readily identifiable product of nitrogen mineralization is formed continuously in the soil and water of a flooded paddy. Organic matter decomposition in absence of oxygen and high pH favours ammonia volatilization from flooded soil.
Ammoniated superphosphate	A product obtained when superphosphate is treated with ammonia or with solution which contains ammonia or other compounds of nitrogen.
Ammoniating solution (nitrogen solution)	A solution used for ammoniating superphosphate or a mixture of superphosphate with other fertilizers. This solution may be liquor ammonia itself or a 'solution of ammonium nitrate or urea in liquor ammonia.
Ammonical nitrogen	Presence of nitrogen in form of ammonia which is denoted as NH_4-N. Ammoniated fertilizers are carriers of ammonical nitrogen. Generally rice plant uses NH_4-N from reduced or submerged conditions.
Ammonification	Production of ammonia as a result of the biological decomposition of organic nitrogen compounds.
Ammonifying bacteria	Bacteria which decompose protein and other nitrogenous substances to produce ammonia.
Ammonium chloride (salt ammonia or NH_4Cl)	Ammonium salt of hydrochloric acid containing 25% of nitrogen in ammonical form.
Ammonium citrate	A term used to express the soluble phosphate content of fertilizers, to describe various extraction solutions of given concentrations of ammonium citrate and aqueous ammonia.
Ammonium diffusion (soil)	Diffusion of NH_4^+-N from the anaerobic soil layer to the aerobic soil layer where it undergoes nitrification and denitrification.
Ammonium fixation	Adsorption of ammonium ions by soils or minerals in such form that they are neither water-soluble nor readily exchangeable.

Term	Terminology
Ammonium nitrate (NH_4NO_3)	The ammonium salt of nitric acid containing 33% nitrogen, half of which is in the ammonium form and the other half in the nitrate form. It is in granular or pilled form.
Ammonium phosphate	A product obtained when phosphoric acid is treated with ammonia, it consists principally of monoammonium phosphate, diammonium phosphate or a mixture of these two salts.
	A material produced by neutralizing a mixture of phosphoric sulphate acid and sulphuric acid with ammonia.
Ammonium sulphate $[(NH_4)_2SO_4]$	Ammonium salt of sulphuric acid containing 20.6% of nitrogen in ammonical form.
Ammonium sulphate nitrate	A double salt of ammonium sulphate and ammonium nitrate containing 26% of nitrogen.
Amphi-photoperiodic plants	Plants which flower either under shorter or longer photoperiods, but not in the intermediate range, for example, *Media elegans* and *Setaria verticillata*.
Amphithermo-photoperiodic plants	Plants in which flower initiation is promoted by relatively short or long photoperiod but not in an intermediate day-length rage. For example certain strains of Media elegans flower on 8 hours and 18-24 hours days.
Anabolism	An energy-consuming process, where in complex organic substances are produced from simple materials.
Anaerobe	An organism that survives in the absence of air or molecular oxygen.
Anaerobic respiration	Incomplete oxidation of foods in living cells with release of energy and does not require atmospheric oxygen. It is also called intramolecular respiration.
Analysis	As applied to fertilizers, it designates the percentage composition of the product expressed in terms which the existing trade practice and law require and permit.

Term	Terminology
Analysis of variance	This is a statistical technique by which source of variation of (ANOVA) any attribute may be partitioned and estimated.
Anemometer	An instrument used in the measurement of force and velocity of wind.
Aneroid barometer	An instrument used for measuring atmospheric pressure. Its main advantage over a mercury barometer is its portability.
Angiosperm	One of the flowering plants whose seeds are enclosed in fruits. A plant having its seeds enclosed in an ovary.
Angstrom	A unit of length equal 10^{-8} cm.
Angular transformation	Before the analysis of variance technique is a field, the data having binomial distribution is transformed to normal distribution by making use of angular transformation; the data relating to percentages are transformed to degrees.
Anhydrous	Dry or without water.
Anhydrous ammonia	Contains 82% nitrogen in liquid form obtained by compressing ammonia gas. It is highly acidic, with 148 equivalent acidity. It requires special equipment to apply to the soil.
Animal day	One day's tenure upon range or pasture by one animal.
Animal unit	One mature cow (454 kg) or the equivalent based upon average daily forage consumption of 12 kg dry matter/day.
Animal unit month (AUM)	The amount of feed or forage required by an animal unit for one month.
Anion	An ion carrying negative charge, for example Cl^-, SO_4^- etc.
Anion or salt respiration	The increase in respiration where a tissue or plant is transferred from water to salt solution.
Anion-exchange capacity	The sum total of exchangeable anions that a soil can absorb, expressed as milliequivalents/100 g of soil (or of other adsorbing material such as clay).

Term	Terminology
Annidation	Complementary use of environmental resources by intercrop components.
Annual	Plant that completes its life-cycle from seed to seed within one year.
	A plant that completes its life cycle in a year or less. It germinates from seed, and grows flowers, produces seed and dies in the same season.
Antagonism	Negative action of chemicals used in combination so that the total action is inferior to their independent effects.
	The action of two or more chemicals in a mixture, with a total effect smaller than the most active component applied alone.
Anthesis	The opening of a flower bud. The duration of life of a flower from the opening of the bud to setting of the fruit.
Antibiotic	An organic compound produced by micro-organisms or higher plants which, in sufficient concentration kills or inhibits growth of certain other organisms.
Anticyclone	A region or an area of high atmospheric pressure in relation to its surroundings diminishing outwards from the centre.
Antidote	A substance intended to counteract the effect of a poison.
Anti-proton	A particle possessing exactly the same properties of proton except that its charge is negative.
Anti-quality constituents	The toxic substances present in forage plants which may either cause direct metabolic damage to the animal or interfere with some phase of digestive utilization are referred to as anti-quality constituents.
Antitranspirants	Substances that retard or inhibit transpiration, *e.g.,* PMA, kaolin etc.
Apatite	A naturally occurring complex calcium phosphate which is the original source of most of the phosphate fertilizers.

Term	Terminology
Apical dominance	Inhibiting effect of a terminal bud upon the development of lateral buds.
Apical meristem	Tissue at the tip of a root or shoot where active cell division occurs.
Apomixis	The development of an embryo from cell other than a fertilized egg. The setting of seed without fertilization as nuclear embryo in citrus.
Apoplast	Non-living cells (xylem) forming a continuous system for movement of water in the plant body or nonliving continuum in plants through intercellular spaces and cell walls.
	The continuous, non-living cell wall structure that surround the simplest, forming a continuous translocation system.
Apparent digestibility	Refers to the balance of feed ingested less that matter lost in faeces, usually expressed as a percentage; obtained by multiplying the digestion coefficient for a nutrient by its content in the feed. Thus the apparent digestibility of ration is the percentage difference between the quantity of food consumed and faeces produced.
Apparent specific gravity (soil)	It is denoted as 'AS' and worked out by following formula: AS = Bulk density of soil/density of water.
Application	General term for all processes of administering fertilizers and soil conditioners to a crop or soil or both.
Appropriate technology	It is a technology that may be suitable or proper in the context of a particular community, region, or country. It must be economically viable, technologically feasible and should fit in the socio-economic fabric of the local communities.
Aquaculture	Farming in water, where fish, algae, shell-fish etc. are grown under controlled conditions.
Aquatic	A plant that grows in water. There are 3 kinds: submergent grows beneath the surface such as waterweed and green algae; emergent, grows above the water, such as cattails and water lilies; floaters; such, as water-hyacinth and water lettuce.

Term	Terminology
Aquatic weed	A weed that grows in water; there are three kinds: (a) submersed weeds, which grow beneath the surface, (b) emerged weeds, which grow roots below but extend above the surface of water and (c) floating weeds, which float on the surface.
Aqueous ammonia	A solution containing water and ammonia, in any proportion, usually qualified by a reference to ammonia vapour pressure.
Aquifer	A water bearing formation in the ground that will yield enough water.
Arable crops	These are crops which require cultivation.
Arable farming	Farming system that involves the production of crops requiring tillage.
Arable land	Cultivated land used for growing crops, *i.e.,* not covered by natural vegetation or permanent grass field.
Arboriculture	Cultivation of woody plants, particularly those used for decoration and shade.
Area-time equivalency ratio (ATER)	It is the ratio of number of hectare-days required in monoculture to the number of hectare-days used in intercropping to produce identical quantities of each of the component crop.
Areole (islet)	A small area of mesophyll delimited by intersecting veins.
Arid climate	A generally extremely dry climate with an annual average precipitation usually less than 250 mnt and generally rainfall is well short of evapo-transpiration demand the atmosphere.
Arid lands	Area where precipitation is 'insufficient for crop production.
Arid zone	A zone of very low rainfall with most of the deserts.
Aridisols	Mineral soils that have an aridic moisture regime, an ochric epipedop, and other pedogenic horizons, but no oxic horizon. These are desert soils with weak ochric.
Aridity	'Refers to excess of evaporation over precipitation.
Aridity index (AI)	A measure of dryness of a region.

Term	Terminology
Arithmetic mean	The average obtained by summing the value of a series of items and dividing by the number of items in the series.
Aroma	The distinctive smell imparted by the volatile constituents present in the planting material, its distilled essential oil and oleoresin extract.
Aromatic	Chemical compound derived from the hydrocarbon benzene (C_6H_6) and having a pungent odour.
Aromatic compound	Carboxylic compound containing a certain amount of unsaturation in the ring.
Aromatics	Compounds derived from the hydrocarbon benzene (C_6H_6).
Arrowing	The flowering in sugarcane is referred as arrowing.
Artificial fertilizers	Fertilizers made by chemical process or mined. Also known as inorganic or mineral fertilizers, chemical fertilizers, artificial manures etc.
Artificial radioactivity	Radioactivity that is acquired as a result of nuclear activity. It denotes the phenomenon of radioactivity produced by particles bombardment or electromagnetic irradiation.
Arviculture	Crop science.
Asexual reproduction	Reproduction of an organism without involving the germ or sexual cells, such as by stolons (vegetative) or by apomixes seed in certain plants.
Ash	The non-volatile residue resulting from complete burning of organic matter.
Aspirated psychrometer	Psychrometer with artificial ventilation of two thermometers (dry bulb and wet bulb) like Asmann psychrometer or whirling psychrometer.
Aspiration	It is the process of cleaning by airblast and separating the foreign material which is substantially lower in specific gravity than the produce to be cleaned.
Assimilation	The process whereby photosynthates are utilized in the building up of protoplasm and cell wall.
Assimilation number	It is the amount of CO_2 absorbed in grams per hour for each gram of chlorophyll during photosynthesis.

Term	Terminology
Assimilatory power or reducing power	It is energy-deriving power of ATP and $NADPH_2$, which is needed for reduction of CO_2 to carbohydrate by plants.
Assimilatory quotient	It is the ratio of the amount of CO_2 produced per each molecule of O_2 consumed.
Atmometer	An instrument used for measurement of evaporation rate, also called atmidometer or evaporimeter.
Atmospheric pressure	The pressure exerted by the atmosphere as a consequence of the weight of the air lying directly above the unit of area in question. At sea-level atmospheric pressure is equal to 76 cm Hg column.
Atom	The smallest unit of an element that can take part in a chemical reaction. It consists of a minute nucleus carrying positive charge, surrounded by negatively charged electrons which revolve round the nucleus in orbit. Nucleus consists of positive particles were emitted by radioactive nuclei, protons and neutral neutrons. Atom as a whole is electrically neutral.
Atomic energy	A misnomer for nuclear energy but accepted because of common usage of denote the energy released in nuclear reactions.
Atomic mass	The mass of a neutral atom expressed in atomic mass units. The atomic mass unit (AMU) is defined as exactly one-twelfth the mass of the carbon isotope.
Atomic number	The number of positive units of electricity contained in the nucleus of the atom is equal to the number of negative charges presented by the electrons, for example, atomic number of hydrogen is 1.
Atomic weight	The average weight of the neutral atoms of an element existing as a mixture of isotopes in the same ratio as found in nature.
ATP (Adenosinetri phosphate)	Adenosine-derived nucleotide that is the primary source of energy for plants through its conversion to ADP.
Attenuation	The decrease in the intensity of radiation caused by the absorption and scattering of the radiation as it passes through the matter.

Term	Terminology
Auger (Soil)	A tool for boring holes into the soil. A soil auger is used to withdraw a small soil sample for observation.
Augmentation cropping	It means growing of a crop to supplement the yield/monetary return accruing from the main crop. For example, growing of forage brassica to augment the fodder yield of berseem in first cut
Aus rice	Derived from the Bengali word 'Ashu' meaning earliness, Aus are a group of periodically fixed maturing rice varieties that are photoperiod-insensitive. This rice is usually direct seeded in March-April with pre-monsoon showers and harvested in July-August, middle of the monsoon season.
Autoecology	The study of the details of how an individual or a species interacts with its environment.
Auto-inhibition	Allelochemicals inhibit more strongly plants of the producer species itself.
Autoradiography	Method for recording the distribution of radioactive material in an object (or its section thereof). Performed by placing the surface of the object in contact with photographic emulsion which by blackening on development indicates where particles were emitted by radioactive nuclei?
Autotroph	An organism which makes food from inorganic raw materials; for example, green plants.
Autotrophic	Capable of self nourishment by using CO_2 and carbonates as the sole source of carbon and a simple inorganic nitrogen compound for metabolic synthesis.
Auxanometer	An instrument which is used to measure growth of plants.
Auxins	A group of growth-regulators that (nay stimulate cell growth, root development and other growth process including seed germination.
Available moisture	It is the moisture range or limit available to plants which lies between field capacity and wilting point of a soil.

Term	Terminology
Available nitrogen	A small proportion of relatively large amount of N present in the rooting zone of soils, which is directly available to plants. It is mainly in the form of NO_3 present in soil solution and the dissolved and adsorbed NH_4.
Available nutrient	A portion of any element or compound in the soil that can be readily absorbed and assimilated by growing plants (available should not be confused with exchangeable).
Available phosphoric acid	It is that part which is soluble in water or in a weak dilute acid, such as 2% citric acid.
Available water	The portion of water in a soil that can be readily absorbed by plant roots considered by most workers to be that water which is held in the soil against a pressure of up to approximately 15 bars moisture which lies between field capacity and wilting point.
A-value technique	Radio-chemical analysis of plants grown on soils which have been treated with fertilizers containing elements such as radio-active phosphorus; may be used to calculate the phosphorus supply of the original soil.
Avenue crops	Such crops are grown along farm roads and fences *e.g.,* Arhar, Glycericidia, Sisal etc.
Average intake rate	Rate of infiltration of water into the soil obtained by dividing the total depth of water in filtrated by the total time, from start to finish, of water application; mm/hr.
Avogadro's number	The· number of molecules in a gram molecular weight of any compound (6.023×10^{23}).
Awn (arista)	Stiff bristle-like appendage occurring frequently on the flowering glumes of grasses and cereals.
Axillary tiller or bud	New shoot or bud growth arising from the junction of leaf and stem.
Azolla	A group of aquatic ferns capable of fixing high level of nitrogen from the atmosphere and are widely grown as a fertilizer crop in lowland rice-cultivation systems.

Term	Terminology
Azospirillum	A nitrogen-fixing root and soil-inhabiting bacterium in tropics.
Azotobacter	An aerobic, non-symbiotic nitrogen fixing bacterium.
B horizon	The subsoil layer in which certain leached substances are deposited.
Back cross	In plant breeding a cross of hybrid with one of its parents and the purpose is to transfer a specific gene from an undesirable variety to another commercially desirable one lacking in that particular character.
Back diffusion	Under some conditions the concentration of certain ions may build up at the root surface because the root is unable to absorb them at a sufficiently rapid rate.
Back furrow	A raised ridge left at centre of the strip of land when ploughing is started from the centre to side.
Background (nuclear)	The counts registered in the counter are not only due to the disintegrations of the sample but also due to those contributed by cosmic radiation, natural radioactivity in the vicinity, artificial radioactivity due to fall out or an X-ray unit nearby thermal noise, etc. These extra counts due to the above factors are called Background counts. Actual counts = Observed counts - Background counts.
Bacteria	A large group of single-celled microscopic organisms widely distributed in the air, water, soil, animal and plant tissues including foods.
Badlands	A land type nearly devoid of vegetation, especially a region where erosion has cut the land into an intricate maze of narrow ravines, sharp crests and pinnacles, resulting from serious and semi-arid erosion of soft geologic materials. Most common in arid regions.
Bagasse	The mill residues from the cane sugar industry consisting of crushed stalks from which the juice has been extracted.
Baking soda	Refers to sodium bicarbonate.
Balanced fertilizer	A soil additive containing suitable proportions of each necessary mineral element to grow a plant or a crop.

Term	Terminology
Balanced rations	Rations for livestock, with proper balance between digestible protein and total digestible nutrients and between concentrates and bulky roughage. In addition, the mineralsaild vitamins should also be in adequate quantities.
Ballast elements	Elements such as aluminium and silicon have been called ballast elements because they are generally present in large amounts, although the plant can grow normally without them.
Bamboo spiels	These are used for conveying irrigation water through small levies or bunds. It is made from hollow· bamboo having 30 cm length and 8 cm diameter.
Band application	An application to a continuous restricted band along a crop row rather than over the entire area, mostly done in case of total killer or non selective herbicides.
	An application to a continuous restricted band (or area) such as in or along a crop row rather than over the entire field area.
Band placement	Fertilizer placement in bands on one side or both the sides of the row about 5 cm below the seed and 5 cm to one side of the seed or plant grown, or around each plant.
Band seeding	Placing the seed in rows directly above, but not in contact with a band of fertilizer.
Band treatment	Applied to a linear restricted strip on or along a crop row rather than continuous over the field area.
Bar chart or bar diagram	The graphical representation of frequencies or magnitudes by rectangles drawn with height proportional to the frequencies or magnitude concerned.
Barnyard	Land immediately adjacent to a barn normally fenced to enclose livestock or fowls.
Barogram	The continuous record of atmospheric pressure made by a barograph.
Barograph	A self recording barometer in which continuous trace of the atmospheric pressure is made on an aerogram.

Term	Terminology
Barometer	An instrument used for measuring the atmospheric pressure.
Barrage	A large structure erected across a river in order to store water, usually for irrigation.
Barren	An area devoid of vegetation.
Basal application	An application to the stems of plants at and just above the ground.
Basal dose	Amount of manures or fertilizer applied to the soil just before the crop is sown or planted or transplanted.
Basal treatment	Applied to encircle the stem of a plant above ground such that foliage content is minimal. A term mostly used to describe treatment of woody plants.
Base (period)	Period in days during which irrigation is supplied to a crop.
Base crop	Base crop or main crop is the one is planted at its optimum sole crop population in an intercropping situation and the second crop is planted in between rows of the main or base.
Base data analysis	Comprises the computation and synthesis of available data for delineating homogeneous areas and for determining research priorities and strategies.
Base exchange capacity	The extent to which the soil can absorb exchangeable cations other than hydrogen and aluminium.
Base saturation	The extent to which the adsorption complex of a soil is saturated with exchangeable cations other than hydrogen and expressed as a percentage of the cation-exchange capacity.
Basic intake rate	Rate at which water will enter the soil when after initial wetting of the soil the rate becomes essentially constant; which is equal to saturated hydraulic conductivity'; mm/hr.
Basic slag	A by-product of steel industry obtained from phosphate iron ores. It contains about 6 – 18% phosphoric acid.

Term	Terminology
Basicity of an acid	The number of replaceable hydrogen atoms present in 1 molecule of an acid HCl, monobasic; H_2SO_4' dibasic; H_3PO_4 tribasic.
Basin irrigation method	In this method, water is applied in small plots or nearly flat plot formed in the field by ridges.
Basin listing	Tillage that forms lister furrows at regular intervals to create small basins to capture and store rain or applied water.
Basipetal	Towards the base of a plant organ; generally downwards in shoots and upwards in roots. Opposite of acropetal.
Basophiles	The plants which can grow in alkali soils (pH range 7.4-8.5); also known as alkali plants.
Bast	The fibrous portion of the phloem or inner bark of a plant.
Baule unit	Unit of fertilizer or any other growth factor is taken as that amount necessary to produce a yield of 50% of the maximum possible.
Beam (plough)	The part which transmits the power of the animal to the plough.
Bearded (awned)	Having an awn or bearing long hair.
Bed	Narrow flat-topped ridge on which crops are grown with a furrow on each side to facilitate irrigation and for drainage of excess water; or area in which seedling or sprouts are grown before transplanting
	A narrow-flat tipped ridge on which crops are grown with a furrow on each side for drainage of excess water or area in which seedlings or sprouts are grown before transplanting.
Bed planting	A method of planting in which the seed is planted on beds. Often two or more seed rows are planted on each bed (See ridge planting).
Bedrock	The solid rock underlying soils and the regolith, or exposed at the surface without a cover.
Beer's law	Beer's law describes the light penetration into a crop canopy if the foliage distribution is uniform in horizontal line.

Term	Terminology
Belt of soil water	The part of the zone of aeration which consists of soil and other materials that lies near enough to the surface to discharge water into the atmosphere in perceptible quantities by the transpiration of plants or by evaporation from the soil.
Benchmark	A point of reference in elevation surveys.
Bench terrace	Constructed to make sloping land cultivable or stable and consists of series of platforms or mostly level benches cut into the hill slope in a steep like formation; the platforms being separated by very steep sides by rock or vegetation.
Benchmark survey	A systematic study aimed at collecting data, *e.g.,* on existing crops, varieties, yields, socio-economic constraints, before a project begins. Data collected depict the existing picture of the survey areas with regard to selected parameters and can be used to evaluate the results of the project.
Benefit: Cost ratio	The present value of benefits divided by the present value of costs.
Bentonite	A natural colloidal hydrated aluminium silicate.
Berm	A narrow strip of land kept between channel section and the bank.
Beta decay	The radioactive disintegration of a nucleus resulting in the emission of an electron (beta particle).
Beta oxidation	It is an important feature of oxidation of fatty acids and its role is well understood in the conversion of the herbicidally – inactive compound (MCPB) to the active one (MCPA).
Beta particle	An electron, either positive or negative. Positive beta particles are called positrons. Negative beta particles are called negatrons. The term beta particle and the symbol are reserved for electrons originating in a nucleus.
Beta-ray	A stream of electrons given off from a radioactive substance.

Term	Terminology
Beta-ray spectrometer	An instrument used to measure the energy distribution of beta particles.
BHC (Benzene hexachloride)	An effective economic chlorinated hydrocarbon insecticide.
Bias	A consistent and, false departure of a statistic from its proper value.
Biennial	A plant that completes its growth in 2 years. The first year it produces leaves and stores food, the second year it produces fruits and seeds.
	A plant that completes its growth in two years; produces leaves and stores food during the first year, and produces fruits and seeds during the second-year.
Bin (storage)	It is an enclosed structure used for storage of seeds.
Binder	A machine for cutting a crop and tying it into bundles with twine.
Binding energy	The energy with which a particle is held to an atom: or nucleus.
Binomial distribution	Also known as Bernoulli distribution is a discrete probability distribution expressing the probability of one set of dichotomous alternatives that is success or failure.
Bioagent	A living organism employed to control a pest.
Bioassay	A test method using living organisms to determine the presence of a chemical quantitatively or qualitatively.
Bioassay (biological assay)	Determination of the relative strength of a substance (hormone, drug etc.) by comparing its effect on a test organism with that of a standard preparation.
Biochemical Antagonism	The type of antagonism that occurs when one agrochemical (the antagonist) decreases the amount of a given herbicide that would otherwise be available to its site of action in the absence of the antagonist.
Bio-climatics	A study which deals with the effects of climatic factors, including geography and elevation upon plant responses.

Term	Terminology
Bioconversion	A general term describing the conversion of one form of energy into another by using biotic sources like animals, plants or micro-organisms. Synthesis of organic compounds from carbon dioxide by plants is bioconversion of solar energy into stored chemical energy.
Biodegradation	Oxidative breakdown of synthetic or natural organic substances by microbial activity.
Bio-energy	Energy from biotic sources.
Biofertilizer	Preparations containing live or latent cells of efficient strains of nitrogen-fixing, phosphate-solubilizing or cellulolytic micro-organisms used for application to seed, soil or composting areas with the objective of increasing the number of such micro-organism and accelerate those microbial processes which augment the availability of nutrients that can be easily assimilated by plants.
Bio-gas	A mixture of gases containing methane, carbon dioxide, hydrogen and traces of few other gases produced by the anaerobic fermentation of easily decomposable cellulosic materials in presence of methane-forming bacteria.
Bio-herbicides	These are native pathogens cultured artificially and sprayed just like post emergence herbicides on foliage of weeds to control them.
Biological control	Controlling a pest (weed, insect nematode or pathogen) by its natural or introduced enemies.
	Control of pests using bio-agents like parasites, predators and pathogens that kill the pest.
Biological determinants of cropping systems	The biological factors such as crop species, varieties, weeds, insect pests, and diseases, which determine the crop configuration and performance of a cropping pattern at a given site.
Biological half-life	The time required for one-half of an administered substance (radio-tracer) to be excreted from the body or from an organ of living tissue.
Biological insecticide	A bio-pathogen like *Bacillus thuringiensis*, which kills insects like a chemical insecticide.

Term	Terminology
Biological magnification	Increase in the concentration of a persistent chemical by the organisms at successive tropical levels a food chain.
Biological mineralization	The conversion of an element occurring in organic compounds to the inorganic form as a result of biological decomposition.
Biological monitoring	Direct measurement of changes in the biological component of a habitat based on evaluation of the number and/or distribution of organisms or species.
Biological value of proteins	The degree of usefulness of a protein in human nutrition.
Biological weathering	Refers to physical and chemical weathering assisted biological agencies.
Biological yield	Refers to total dry matter produced by a plant or a crop in a unit land area.
Biology	A branch of science which deals with living organisms ('Bios' means life; 'logos' means discourse or study).
Biomass	Any organic matter which is available on a renewable basis, including agricultural crops and agricultural wastes and residues, wood and wood wastes and residues, animal wastes, mineral wastes and aquatic plants, usually expressed as dry weight per unit area.
Biome	A major ecological community of organisms maintained under a particular climate zone.
Biometeorology	The science of relationship of life to weather.
Biometry	The branch of science which deals with statistical procedures in biology.
Biophysics	The study of phenomena of living organisms by physical methods; the study of physical phenomena exhibited by living organisms or parts thereof.
Biosphere	That part of the earth does envelop in which living organisms exist in their natural state.
Biostatic effect	Reduced microbial decomposition of a pesticide in the presence of some chemical, called extender. Ego boron is biostatic to 2, 4-D.

Term	Terminology
Biota	Animal and plant life of a particular region considered as a total ecological entity.
Biotechnology	Use of biological processes on an industrial scale to produce materials for use by mankind in daily life.
	The application of biological organisms, systems or processes to manufacturing and service industries.
Biotic	Pertaining to life or to living organism.
Biotic factors	Living components of environment which include micro-organisms, plants of the same or of other species and animals including man.
Biotic potential	The inherent capability of an organism to increase in numbers under ideal conditions.
Biotype	A population within a species that has a distinct genetic variation.
Bird scarer	A mechanical device used for scaring away birds; produce loud or bursting sounds at regular intervals.
Biroype	A population within a species that has a distinct genetic variation.
Bisexual	Having both male and female reproductive organs.
Biuret	A condensation product of 2 molecules of urea and is toxic to many plants. Its content in the commercial fertilizer should not exceed 2%.
Black cotton soil	Soil of black colour (throughout the profile) with low organic matter and a high clay content, exhibiting a strong phenomenon of swelling and shrinkage; during dry periods cracks develop that maybe several cm wide and more than 50 cm deep.
Blade	The flat portion of a grass leaf above the sheath
Blanching	To prevent from becoming green, by excluding light, as lettuce leaves or celery.
Blanket application	An application of spray or granules over an entire area rather than only in rows, pockets or as band.
	An application (also called broadcast application) of spray or granules over an entire soil surface or weed-infested area rather than only on rows beds or middles.

Term	Terminology
Blending	The process of mixing two or more different products together to obtain desired final product.
Blind cultivation	Cultivating with a harrow weeder, rotary weeder or other implements to kill weeds before a seeded or planted crop has come up.
	Cultivation of the field before the crop plants emerge.
Bloat	Excessive accumulation of gases in the rumen of the animals. It is due to the formation of stable foam within the rumen which traps the fermentation gases. The main foaming agent is now known to be a plant cytoplasmic protein with a sedimentary velocity of 18 svedberg units (188) also called as Fraction-I. It occurs to the extent of 4-5% in bloat producing forages such as Lucerne but its concentration is < 1 % in non-toxic forages.
Blood meal	Liquid blood is dried with hot air or steam. About 20-25 kg of dried blood can be had from 1,000 kg of liquid blood. It contains 10-12% N, 1-2% P_2O_5 and about 1% K_2O. It is manure.
Blotch	A disease showing large and irregular spots on stems, shoots and leaves.
Blue-green algae	A heterogeneous group of prokaryotic photosynthetic nitrogen-fixing organisms which contain chlorophyll 'a'. They include unicellular, colonial and filamentous species. They are obligating phototrophs and store cyanophycean starch.
Bog soils	An intra-zonal group of soils with a muck or peaty surface underlain by peat; developed under swamp or marsh type of vegetation; mostly found in humid or sub humid regions.
Bolting or Shooting	Refers to significant stem elongation that precedes flowering in many plants.
Bombardment (nuclear)	The act of subjecting a substance to a flux of neutrons or other high-energy particles.
Bone ash	It is obtained by burning bones with free access to air and contains 30-40% P_2O_5.

Term	Terminology
Bone black (bone, char)	A product obtained by heating bones in closed retort. This is used for cleaning sugarcane juice and the spent material used as phosphorus fertilizer. It may contain 1 2% nitrogen and 30 – 35% P_2O_5.
Bone meal (raw)	A fertilizer made of dried animal bones finely ground. It contains 20-24% P_2O_5 of which 16% is citrate soluble. The availability of its plant food depends largely upon how fine it is ground. It contains at least 3% N also.
Bone meal (steamed)	A product made from grinding bones previously treated with steam under pressure. It contains 1-2% nitrogen and 28% P_2O_5 of which 16% is citrate soluble.
Bonus crop	An additional crop of economic importance taken in intercropping or mixed cropping systems without any extra inputs.
Boot stage	Growth stage of grasses at the time the head is enclosed by the sheath of the uppermost leaf.
Booting	State of plant growth indicated by swelling of the flag leaf.
Borax	A chemical compound that contains approximately 11% boron. Borax is applied to boron-deficients pil or sprayed on the plant's foliage which indicates boron deficiency.
Border crops	Such crops help to protect other crops from trespassing of animals or resist the speed of weed and are mainly grown as border.
Border irrigation (border-strip irrigation)	It is an efficient method of irrigating close growing crops and consists essentially of dividing the field by low flat levels into series of strips, each of which is flooded separately. Generally, these strips extend in the direction of the steepest slope (not more than 3%) and, at right angles to the supply ditch. These strips are made level transversely but follow a gentle down field slope longitudinally.
Border rows	These are outer few crop rows that are used to isolate the main plot. In case of seed plots these are the rows of male sterile parents but in case of commercial crops the same seed material is used for raising the border rows.

Term	Terminology
Boro rice	Winter-season rice transplanted in December-January and harvest at April-May. Brown rice is photoperiod-insensitive and grown under irrigated conditions.
Boron toxicity	Symptoms are the appearance of yellow discolouration of the leaf tips that spreads along the margins. Large brown eliptkal spots appear along the leaf margins. Affected parts turn, brown and wither. Boron toxicity occurs on coastal soil arid region soils, soils irrigated with high, boron water and geothermal areas.
Botanical composition	Botanical composition of a pasture refers to the relative proportions of component species and can be expressed on the basis of weight, number. Of individuals, frequency of occurrence or area covered. Proportion of species on weight basis is generally the most useful measure where the main interest is in pasture production. Botanical composition in terms of number, frequency and area is concerned mainly with the pasture as an association of species and individual plants. Such measurements can be used to measure persistence and changes in population due to treatments or time.
Botanical pesticide	A pesticide obtained .from plants. Also called plant-derived pesticide.
Botany	A branch of biology which deals with the study of plants.
Bound water	Form of water that is adsorbed strongly by colloids. This water is retained against forces of evaporation over 1,000 atmosphere.
Bouyoucos moisture meter	Equipment used for the soil moisture in situ with the help of gypsum block or nylon block fixed in field. It was invented by the scientist Bouyoucos.
Bowen ratio	The ratio of energy flux upward as sensible heat to latent energy flux in the same direction (negative when the fluxes are in opposite directions).
Brackish	Slightly saline.
Bran	The pericarp, testa and the aleurone layer of cereal seeds which are removed in milling.

Term	Terminology
Brand	The name, number, trade mark or designation applied to a chemical product of any particular description by the manufacturer, distributor, importer, or vender thereof.
Bray's nutrient mobility concept	As the mobility of a nutrient in the soil decreases, the amount of that nutrient needed in the soil to produce a maximum yield (the soil nutrient requirement) increases.
Bread wheat (common wheat)	Most commonly grown wheat, a hexaploid of the group *Triticum aestivum.*
Break crop	A crop grown in rotation to break continuity of some crop-bound or crop associated weed.
Breeder seed	It is the seed directly controlled by the originating or sponsoring plant institution or individual, and the source of the production of seed for the certified classes. It is genetically purest seed stock.
Bremstralung	Radiation produced when an electron passes through the field of an atom or nucleus, also called breaking radiations.
Brix	The percentage of total solids in sugarcane juice, read from brixhydrometer.
Broad bed and furrow (B.B.F.)	A land management system for successful crop production in system vertisols (heavy black soil) by providing better drainage through raised broad beds (about 90-100 cm wide) and furrow (25-30 cm wide).
Broad-base terrace	A ridge-type terrace 25-50 cm high and 5-10 m wide with gently sloping sides, a rounded crown and a dish shaped channel along the upper side constructed to control erosion by diverting run-off along the contour at a non-scouring velocity.
Broadcast	The process of scattering of agricultural inputs, such as seed, fertilizer and manure, on the surface of the soil by hand or by implements to provide uniform distribution of material over the entire field.
Broadcast treatment	Applied as a continuous sheet over the entire treated field.

Term	Terminology
Broadleaf (weeds)	Used in weed terminology to designate a broad group of non-grass like plants, usually dicots.
Brown soils	A group of soils formed in semi-arid climates having a brown surface soil and zone of carbonate accumulation at a depth usually of 25 to 50 cm.
Browse	The part of leaf and twig growth of shrubs, woody vines and trees available for animal consumption.
Brush control	The control of woody-type plants.
Bud	A plant structure that contains an undeveloped shoot or flower.
Buffer compounds (soil)	The clay, organic matter and such compounds as carbonates and phosphates, which enable the soil to resist appreciable change in pH value.
Buffer solution	A solution which resists change in pH by the addition or loss of hydrogen or hydroxyl ions.
Buffer strips	Contour strips of grass or other erosion resisting vegetation between or below cultivated strips or fields.
Bulb	An underground stem with fleshy, food-storing scale leaves, essentially a below ground bud.
Bulk density	It is the ratio of the mass of a soil's solids (oven dried soil) to the bulk volume of the soil. It is expressed as mass per unit volume, generally g/cc.
Bulk fertilizer	Commercial fertilizer delivered. to the purchaser, either in the solid or in the liquid state, in a non-packed form.
Bulk sample	The entire quantity of the sample received in the seed laboratory for testing.
Bulk specific gravity	The ratio of the bulk density of a soil to the mass of unit volume of water.
Bulk volume	The volume of an arbitrary soil mass including the volume of the solid particles and of the pores (interstices, voids).
Bulky organic manures	These manures are bulky in nature and supply plant nutrients in small quantities and organic matter in large quantities.

Term	Terminology
Bulldozing	The pushing and rolling of soil by an inclined blade.
Bund	An artificial earthen embankment made across sloping agricultural land to cut short lengthy soil slopes and reduce run-off and erosion.
Bund former	An implement used for making bunds or ridges by collecting the soil. Bunds are required to hold water in the soil, thereby conserve moisture and prevent run-off.
Buoyancy	The upward force excreted on the volume of fluid by virtue of the density difference between the volume of fluid and that of surrounding fluid.
Bush fallowing	It is an intensive fallow system with bush as the fallow vegetation.
Bush-fallow system	Farming systems in which the natural regeneration of self-propagated plants in succession community is the restorative agent in respect of nutrients, organic matter, water conservation and microclimate. Crop plants are grown on patches cleared by felling of trees and/or burning of grasses and herbs, fallow succession dominated by woody shrubs and grasses.
C horizon	A mineral horizon generally beneath the solum that is relatively unaffected by biological activity and pedogenesis and is lacking in properties diagnostic of an 'A' or 'B' horizon.
C_3 Plant	A plant in which the first product of CO_2 fixation is the 3-carbon compound; phosphoglyceric acid and is comparatively photosynthetically less efficient than C_4 plants, *e.g.,* wheat, rice, barely etc.
C_4 Plant	A plant in which the first product of CO_2 fixation is the 4-carbon compound oxaloacetic acid and is photosynthetically more efficient than C_3 plants, *e.g.,* Sugarcane, maize etc.
Caffeine	It is the purine base alkaloid found in tea and coffee which works as a stimulant.
Caking	The formation of a coherent mass from individual particles.
Calcareous	Composed of or containing calcium carbonate ($CaCO_3$).

Term	Terminology
Calcareous soil	Soil containing sufficient calcium carbonate (often with magnesium carbonate) to effervescence visibly when treated with cold 0.1 N hydrochloric acid.
Calcification	Process including accumulation of calcium carbonate.
Calciphytes	Plant that require considerable amounts of calcium.
Calcium ammonium nitrate (CAN)	A commercial nitrogenous fertilizer, consisting of ammonium nitrate and powdered limestone or dolomite containing 20.0 per cent nitrogen. Onehalf of the nitrogen is in the nitrate form and the remaining half in the ammoniacal form.
Calcium cyanamide $(CaCN_2)$	An organic compound containing around 21 per cent N. It is used as a fertilizer, defoliant and herbicide. It decomposes in soil to yield urea and water.
Calcium metaphosphate $[Ca(PO_3)_2]$	A product obtained by treating phosphate rock with gaseous phosphorus pentoxide (PPs) at high temperature.
Calcium nitrate $[Ca(NO_3)_2]$	The calcium salt of nitric acid. It is an excellent source of the nitrate form of nitrogen and of water soluble calcium. The commercial product contains about 15 per cent nitrogen and 28 per cent CaO.
Calendar of operations	A graphic or tabular presentation showing the kinds of operations to be performed during the season and specified time limits within which work must be done.
Calibration	Series of operations to determine the amount of solution applied per unit area of land and the amount of chemical required for a known volume of diluents.
Caliche	A more or less cemented deposit of calcium-carbonate often mixed with magnesium carbonate at various depths, characteristic of many of the semi-arid and arid soils.
Calorie	A unit of measurement of heat energy and is defined as the amount of heat energy required to raise the temperature of one gram of water through 1°C.
Calorie index	A single productivity index which incorporates caloric equivalents by all crops in a system.

Term	Terminology
Calorimeter	An apparatus used for measuring the amount of heat generated or emitted by any substance.
Calutron	A device used to separate isotopes. It is based on the principle of the mass spectrometer.
CAM plants	The CAM (Crassulacean acid metabolism) system is prevalent in desert plants where the CO_2 is fixed at night. In these species, there is diurnal fluctuation of acidity in thickened leaves. The CAM plants are adapted to environments of more or less constant aridity, *e.g., Bryophyllum calycinum* and cactus plants.
Cambium	A meristem with products of divisions arranged in an orderly fashion in parallel files. It consists of one layer of initial cells and their derivatives.
Cane sugar	A type of storage sugar (sucrose) present in sugarcane, sugar beet, ripe fruits etc.
Canopy	The branches, leaves etc. formed by crops, some distance above the ground: Near about a continuous stratum of foliage.
Canvas dam	It is a water-control device having a triangular piece of canvas cloth fitted with a bamboo stick on one side and used for obstructing flow of water in channels.
Capacity adaptation	A type of adaptation which enables the plant to grow and develop under environmental conditions not permitting normal growth and development of unadapted plants.
Capacity factor	Is the ratio between actual supply to the capacity of a channel or a canal?
Capillary conductivity	Is the distance travelled by soil moisture in a unit time under unit potential gradient in unsaturated conditions?
Capillary fringe	Is the vertical distance above a water table along which moisture content varies from full saturation to field capacity?
Capillary porosity	The volume of small pores within the soil that holds water against the force of gravity.

Term	Terminology
Capillary potential	A measure of the attractive forces with which water is held by soil. It is usually expressed in terms of work that must be done to move water against the capillary forces of the soil.
Capillary rise	The rise of water above the hydrostatic surface through the influence of capillarity.
Capillary water	Is that which is held by the surface tension forces as a continuous film around the soil particles or water retained within soil pores?
Carbohydrates	Compounds of carbon, hydrogen and oxygen in the ratio of one atom each of carbon and oxygen to two of hydrogen, as in sugar, starch and cellulose.
Carbon cycle	The sequence of transformation undergone by carbon utilized by organism wherein it is used by one organism, later liberated upon the death and decomposition of the organism, and returned to its original state to be reused by another organism.
Carbonate accumulation zone	A visible zone of accumulated calcium and magnesium carbonate at the depth to which moisture usually penetrates in semi-arid soils before they are brought under cultivation.
Carbon-nitrogen ratio (C:N ratio)	The ratio of the weight of organic carbon (C) to the weight of total nitrogen (N) in a soil or in organic material.
Carcinogenic	Capable of causing cancer in animals.
Cardinal temperatures	Below a certain minimum and beyond maximum value of temperatures, plant growth ceases. Between these limits, there is an optimum temperature at which growth proceeds with greatest rapidity. These three points are known as cardinal temperatures.
Carotene	An orange pigment occurring in certain plastids and is the precursor of vitamin A.
Carotenoid pigments	Accessory pigments located in chloroplasts as water insoluble protein complexes. These are lipid compounds and range in colour from yellow to purple.

Term	Terminology
Carrier	The liquid or solid material added to a chemical compound to facilitate its storage, shipment and application.
Carrier (radio tracer)	A ponderable amount of stable element mixed with a radio isotopic tracer of that element.
Carrier concept	Refers to the movement of ions across the cell membrane by specific carriers (hypothetical) with the help of metabolic energy.
Carrier free	A radioisotope free of any stable isotope of the same element.
Carry over soil moisture	The moisture stored in crop root zone between cropping seasons or before the crop is planted. This moisture contributes to the water needs of succeeding crops.
Carrying capacity	The maximum number of organisms/animals/plants that can be supported in a given area or habitat over period of year (generally refers to the number of grazing livestock that can be maintained on a given grassland over a period of time).
Caryopsis	A one-seeded dry indehiscent fruit in which the seed coat and pericarp are united, *e.g.,* fruit of cereals and grasses.
Cash crop	A high value marketable crop *e.g.,* sugarcane, cotton, jute, tobacco, tea etc.
Catabolism	Metabolic process in which complex materials are converted into simple compounds, *e.g.,* digestion and respiration.
Catalyst	Substance that accelerates the rate of chemical reaction without being used up in the process.
Catch crop	A quick growing crop incidentally planted and harvested between two major crops in consecutive seasons, also it may be a contingency crop grown to replace a major crop which has failed.
Catchment	An independent unit of treated or untreated land area contributing runoff water to a common reservoir or stream.

Term	Terminology
Catchment area	An area from where entire rain water is collected and then conducted to a reservoir.
Catena	A sequence of soil of about the same age, derived from similar parent material, and occurring under similar climatic conditions but having different characteristics due to variation in relief and in drainage.
Cathode	It is the negative electrode to which the cations are attracted.
Cation	An ion carrying positive charge of electricity. The common soil cations are calcium, magnesium, sodium, potassium and hydrogen.
Cation exchange	The exchange of cations held by the soil-adsorbing complex with other cations. Thus if a soil-adsorbing complex is rich in sodium (as is the case in alkali and alkaline soil), application of gypsum (calcium sulphate) causes calcium cations to exchange with sodium cations.
Cation–exchange capacity	The sum total exchangeable cation adsorbed by a soil, expressed inmilliequivalents per 100 g of soil. Measured values of cation-exchange capacity depend somewhat on the method used for the determination.
Cay mineral	Small mineral soil particles less than 0.002 mm in diameter.
Cay pan	A dense and heavy soil horizon underlying the upper part of the soil profile, hard when dry and stiff or plastic when wet, more or less impermeable to roots.
Ceiling leaf area index	Leaf area index which mutual shading of leaves in a canopy is complete and net photosynthesis is zero i.e. gross photosynthesis equals the respiration losses of the whole crop.
Cell	The structural and physiological unit of plants and animals, generally consisting of cytoplasm and nucleus and also in plant cells, a cell wall.
Cell constituents	Readily available soluble fraction of the forage-dry matter. This corresponds to cellular contents and is composed of lipids, soluble carbohydrates, most protein, non-protein nitrogen and other water soluble. These are completely digestible by animals without the aid of fermentation and are unaffected by the degree of ignifications.

Term	Terminology
Cellulose	A linear homopolysaccharide composed of D-glucopyranoside units linked 1-4.
Cemented soil	A soil in which the grains or aggregates adhere firmly and are bound together by some material that acts as a cementing agents such as colloidal clay, iron, silica, or alumina hydrates, lime etc.
Centre of origin	An area where the given plant species show the greatest genetic diversity.
Cereal forage	A cereal crop harvested when immature for either hay, silage, green feed or as pasturage.
Cereals (crops)	Cereals are the crop plants belonging to the grass family Gramineae which are grown for their edible starchy seed botanically known as caryopsis. The word 'ceres' means the godess of grain, *e.g.,* wheat, rice, maize.
Certified samples	Samples which are submitted by a seed certification officer for the purpose of determining if the lot of seed from which sample is collected is of satisfactory germination and purity to be tagged and sold as certified seed.
Certified seed	Is the progeny of breeder, foundation or registered seed, which is handled to maintain satisfactory genetic purity and identity in a manner acceptable to certifying agency?
Chaff	Glumes, husks or other seed covering combined with other plant parts separated from seed in threshing or processing.
Chaffing	Pneumatic separation of very light materials from the produce.
Char	A porous, solid residue resulting from the incomplete combustion of biomass. If produced from coal, it is called coke, if produced from wood or bone, it is called charcoal. It is closer to pure carbon than the coal, wood or bone from which it is produced.
Character	An identifiable hereditary property, such as a specific component colour, a structural detail or resistance to pests and diseases etc.

Term	Terminology
Characteristics curve or moisture release curve	Is the relationship (function) between soil moisture tension and the moisture content in the range from field capacity to wilting point?
Chasmophyte	A plant whose roots are capable of penetrating into rock fissures.
Check Dam	A small, low dam constructed in a gully or other water course to decrease the velocity of stream flow, for minimizing channel scour and promoting the deposition of eroded material.
Check irrigation (check basin irrigation)	Irrigation method whereby earth ridges are built around that area partly or wholly to retain water to form a pond.
Check row planting	The process of planting in which row to row and plant to plant distances are uniform and plants across the rows are also in line.
Chelate	An organic compound capable of holding the plant nutrient in a form which prevents it from getting tied with other elements in the soil, thus keeping it more or less in available form for the plant. The term refers to the claws of a crab illustrative of the way in which the atom is held, *e.g.,* EDTA.
Chemical Antagonism	The type of antagonism that occurs when an antagonist reacts chemically with an herbicide to form an inactive complex.
Chemical fallow	The substitution of herbicides to tillage tools to control weeds on fallowed land.
Chemical fallowing	Chemical fallowing is the use of herbicides instead of tillage to kill volunteer vegetation on the fallow fields. It leaves the ground under dry vegetation mulch which protects the soil against erosive agencies and paves the way to larger storage of rain in soil.
Chemical name	Scientific name for the active ingredient in a pesticide.

Term	Terminology
Chemisorbed phosphorus	A chemically precipitated monolayer on the surface of another crystalline species, formed by reaction of phosphorus compounds with solid-phase constituents. A form of sorbed phosphorus, involving surface valence forces and chemical affinities between the phosphate ion and the lattice constituents on which sorption takes place.
Chemosynthesis	Food manufacture carried on by certain bacteria utilizing energy released in the oxidation of inorganic compounds.
Chemotropism	A growth movement in response to a chemical.
Chemozem soil	A dark to nearly black grassland soil high in organic matter developed in a subhumid climate.
Chestnut soil	A soil having a dark-brown surface developed under-mixed tall and short grasses in a subhumid to semi-arid climate.
Chickpea	A legume (grain) crop also named as gram, Bengal gram or Kabuli gram grown in winter as pulse crop.
Chilean nitrate of soda	A product obtained by refining the crude nitrate deposits found in Chile and containing about 99 per cent sodium nitrate.
Chilling injury	An injury caused by relatively low temperatures above the freezing point.
Chips	Bark obtained from thick branches and stems.
Chisel	A tillage implement with points about a foot apart that stir the soil to a depth of 25 to 45 cm.
Chiseling	Breaking or loosening compact soil or subsoil with a chisel cultivator, in order to improve the root penetration and absorption of water.
Chi-square (χ^2) test of independence	Test to determine whether two attributes are independent by comparison of observed frequencies relative to expected frequencies.
Chlorofluoro carbons (CFC)	Compounds of carbon and halogens, released into troposphere, where they dissociate to release free chlorine that causes depletion of ozone.
Chlorophyll	A light trapping green pigment essential for photosynthesis.

Term	Terminology
Chloroplast	A cellular structure (plastid) made up of protoplasm and containing the green pigment chlorophyll wherein the photosynthesis is carried out.
Chlorosis	Yellowing or blanching of the green plant parts due to sub-normal quantity floss of chlorophyll in the tissue.
	The loss of green colour (yellowing) in plant foliage resulting from the impairment or stoppage of the green colouring matter.
Chopping	The process of cutting plant material with knives or other sharp instruments.
Chromoplast	A coloured plastid containing pigments other than chlorophyll, often they are yellowish or red in colour.
Chronic toxicity	The quantity or potential of a substance to cause injury or illness after repeated exposure to small doses over an extended period of time.
Citrate soluble phosphoric acid	That part of the total phosphoric acid in a fertilizer that is insoluble in water but soluble in a neutral solution of citrate of ammonia.
Citric acid cycle (Kreb's cycle)	A cyclic series of reactions which transforms pyruvic acid into carbon dioxide and hydrogen which then goes terminal oxidation to yield available energy.
Citric acid-soluble P_2O_5	That part of the total P_2O_5 particularly in basic slag and bone meal that is insoluble in water but soluble in 2% citric acid solution.
Class' A' pan (USWB evaporimeter)	It is a standard evaporimeter (120.7 cm in diameter and 25 cm deep) maintained in agro-meteorological observatories which is helpful in generating data on evapo-transpiration, which in turn, can be used for scheduling of irrigation.
Clay (soil)	Soil material that contains 40% or more clay, less than 45% sand, and less than 40% silt.
Clay spiles	Used for water control, made up from Chinese clay hollow piece of pipe, having diameter of 8 cm and length of 30 cm or more which are used for conveying irrigation water through small levies or bunds.

Term	Terminology
Clayey	Includes all clay textural classes, i.e., sandy clay, silty clay and clay.
Clean cultivation	Periodic soil tillage to eliminate all vegetation other than the crop being grown.
Cleaning (seed)	The removal of foreign or dissimilar material by washing, screening, hand-picking, aspiration or any other mechanical means.
Cleaning crop	Whose agronomical practices makes the field clean.
Climate	The aggregate of weather conditions over a long period of time.
Climatic index	A number which condenses climatic data into a simplified expression (Transeau's ratio: PIE is a climatic index).
Climatological-normals	Period averages of climatic data calculated over' a uniform and relatively long period covering at least 3 consecutive 10-year periods.
Climatology	The study of weather patterns over time and space. It concerns with the integration of day-to-day weather over a period of time.
Climax	A plant community of most advanced type capable of development under, and in dynamic equilibrium with the prevailing environment.
Climax community	The final or stable state of a community which is in complete harmony with the physical and biotic environment of habitat and is self-perpetuating.
Climax vegetation	The most fully developed natural vegetation the climate can sustain.
Climbing roots	Modified roots of climbers such as betelvine or black pepper which produce roots from the nodes by which they cling to their support and climb it.
Climograph	A graphical representation of the differentiation between various types of climate.
Clod	A compact, coherent mass of soil ranging in size from 5 to 10 mm to as much as 200 to 250 mm, produced artificially usually by the activity of man in ploughing, digging, etc., especially when these operations are performed on soils that are either too wet or too dry for normal tillage operations.

Term	Terminology
Clone	A group of individuals of common ancestry which have been propagated vegetatively, usually by cuttings or natural multiplication, *e.g.,* bulbs or tubers.
Closed formula mixed	The fertilizer grade is disclosed on each bag of such fertilizer mixture but the ingredients or straight fertilizers used in formulating the mixture are not disclosed. In India, fertilizer mixtures sold to the cultivators are usually of the closed formula type.
Clothesline effect	Horizontal heat transfer from a zone of warm upwind over a relatively cool irrigated crop field.
Cloud	Visible aggregate of minute particles of liquid water or ice or both together in suspension in the atmosphere.
Cloud seeding	Injecting the clouds with a seeding agent like dry ice, silver iodide, sodium chloride from an aircraft or using a ground generator for producing artificial rain.
Cloudburst	Storm or rain of extraordinary intensity and of relatively short duration, usually over a small area.
Cloudiness	Degree of cloud cover, usually mean of several observations per day; expressed in octads (in eights) of sky covered, or in tenths of sky covered.
Clubbing	The process of removing stumps, sunken stones, boulders and all roots over a minimum size and up to a specified depth.
Coarse texture	Includes the sands, loamy sands, and sandy loams, except the very fine sandy-loam textural classes. Sometimes subdivided into sandy and moderately coarse textured.
Coastal alluvium	These soils are subjected to the action of sea water, saline, sandy loam in texture, excessive in drainage, low water holding capacity, poor in plant nutrients.
Coated fertilizer	Fertilizer, the granules of which are covered with a thin layer of different materials in order to improve the behaviour and/or to modify the characteristics of the fertilizer.
Coefficient of variation	The standard deviation expressed as a percentage of mean. CV% = Standard deviation ÷ Mean × 100.

Term	Terminology
Cohesion	The force of attraction that binds the molecules of the same kind.
Cold hardiness	A process by which a plant acquires hardiness to withstand cold conditions due to continuous exposure to low temperatures.
Cold storage	An insulated storage utilizing mechanical refrigeration to maintain a stable cold temperature for long-term storage of agricultural products.
Cold test	A germination test in which seeds are planted for a period in cool, moist, unsterilized soil before transfer to a high temperature, designed to approximate possible unfavourable planting conditions in order to determine seedling vigour.
Cole crop	Cole is derived from colewart. Colewart is the ancestor of wild cabbage. Therefore, Cole crops are essential cold weather crops belonging to the conciferae capable for withstands considerable frost.
Coleoptile	A protective sheath covering the first leaf of a grass seedling as it emerges from the soil.
Colloid	A fine particle usually 10^{-6} to 10^{-4} mm in diameter which carries an electric charge. A wet mass of colloidal particles is glue-like in consistency.
Colloidal clay	The smallest fraction particle in a clay fraction; 1 micron or less in diameter, which determines many of the physical and chemical properties of the soil.
Colluvium	Deposit of rock fragments and soil material accumulated at the base of steep slopes by gravitational action.
Columnar soil structure	Similar to prismatic structure except that the tops of the blocks are rounded.
Combined tillage	Operations simultaneously utilizing two or more different types of tillage tools or implements (subsoilerlister, lister-planter or plough-planter combinations) to simplify control or reduce the number of operations over a field.
Combine-harvester-thresher	A machine designed for harvesting, threshing, separating, cleaning and collecting grain while moving through the standing crop in one operation.

Term	Terminology
Command area	The irrigable land area by the canal or from a dam or an irrigation project.
Commensalistic polyculture	The interaction between two crop species has a positive net effect on one species and no observable effect (negative/positive) on the other species, when grown together.
Commercial cane sugar	The convertible sucrose or cane sugar content of sugarcane.
Commercial cash crop	It involves growing crops primarily for sale and not for home production consumption.
Commercial cash energy or non-renewable source of energy	Energy from fossil sources such as coal and petroleum.
Commercial crop	Such crops are grown to earn money.
Commercial farming	The type of farming where capital input is high and the production is market and profit oriented.
Common name	Common name means the name assigned to a pesticide active ingredient by the International Standards Organization or adopted by national standards authorities to be used as a generic or non-proprietary name for that particular active ingredient only.
Community forestry	It is the forest of the community by the community and for the community. Such forestry is managed for the interest of local community or a village and is run almost exclusively by them.
Compaction	The increase in density of a soil as a result of applied pressure or load.
Companion crop	Any subsidiary crop grown in association with a main crop.
Compatibility	Mixable in the formulation or in the spray tank for application in the same carrier without undesirably altering the characteristics or effects of the individual components.
Compatibility	The nature of two compounds that permits them to be mixed without affecting the properties of either.

Term	*Terminology*
Compensation point (light)	It is the light intensity at which the photosynthetic intake of carbon dioxide is equal to the respiratory output of carbon dioxide.
Competition	The tendency of the plants of the same or different species growing together to strive for and capture the common resources, such as light, water, nutrients and space of the habitat. The competition amongst individuals of same species is termed as intraspecific competition, *e.g.,* inter and intra row competition among crop plants. Competition amongst individuals of different species is termed as interaspecific such as crop-weed competition.
	The active acquisition of limited resources by an organism that results in a reduced supply and consequently reduced growth of other organisms in an environment.
Competition effect	Competition of inter-cropped species for light, nutrients, water, carbon dioxide and other growth factors.
Competition index	It is measure of competition based on product of two equivalence factors, one for each component species in mixed stands, where equivalence factor denotes the number of plants of one species which is as competitive as one plant of the other species. If a given species has an equivalence factor of less than I, it means it is more competitive (on a plant for-plant basis) than the other species. If the competition index is less than I, there has been an advantage of mixing.
Competitive Antagonism	The type of antagonism that occurs when an antagonist acts reversibly at the same site as the herbicide.
Competitive crops	Such crops compete to each other and are not suitable for intercropping.
Competitive enterprises	The enterprises which compete for the same resource at the same time or season.
Complementary effect	Effect of one component on another which enhance growth and productivity, as compared to competition above.

Term	*Terminology*
Complementary enterprises	The enterprises which do not compete for resources such as land, labour, capital and management but mutually contribute to the production of each other.
Complementary intensive cropping	Growing of morphologically and physiologically different crops in an agricultural year under the sequential intercropping systems, so as to produce complementary effects upon each other and on succeeding associations.
Complete fertilizer	A product obtained by mixing different fertilizer stock materials, containing three major plant nutrients, namely, nitrogen, phosphorus and potassium (See also Fertilizer Mixture).
	The commercial fertilizers containing at least two or more of the major nutrients, made by a chemical reaction/between the nutrient containing raw materials.
Complimentary ion effect	The influence of one adsorbed ion on the release of another from the surface of a colloid.
Component crops	Individual crops making up a cropping system.
	Individual crop species which are a pmt of the multiple cropping system.
Component technology	The cultural techniques used in the management of a crop or cropping pattern. Component technologies include variety, planting method, tillage operations, fertilizer management, pest management, harvesting etc.
	The procedure for growing each component crop.
Composite sample	A seed sample composed by mixing different sub-samples taken from various parts of a seed lot.
Compost	A product obtained by the controlled decomposition of organic wastes, finally used as organic matter.
Composting	A biological process in which micro-organisms of both types, namely, aerobic and anaerobic, decomposes the organic matter and lowers the carbon: nitrogen ratio of the refuse. The final product of composting is well-rotted manure known as compost.

Term	Terminology
Compound fertilizer	Fertilizer having a declarable content of at least two of the nutrients nitrogen, phosphorus and potassium, obtained chemically or by mixing, or both.
Compound land-utilization type	A land-utilization type consisting of more than one kind of use or purpose, either undertaken in regular succession on the same land, or simultaneously undertaken on separate areas of land which for purposes of evaluation are treated as a single unit.
Compton effect	The change in frequency of wavelength of photons when they are scattered by electrons and nucleons.
Concentrate (pesticides)	A pesticide, as sold, before being diluted for application.
Concentrates (feed)	Fortified supplement feed which contains all necessary ingredients to form a complete balanced ration for animal, fowl or fish.
Concentration	Amount of active ingredient in a given volume or weight of a formulation or mixture.
	The amount of active ingredient in a given volume of diluents or given weight of dry material.
Concretion (soil)	Hardened, local concentrations of certain chemical compounds, such as calcium carbonate or iron oxides, in the form of indurate grains or nodules of various sizes, shapes and colours.
Conditioner (of fertilizer)	A material added to a fertilizer to prevent caking and to keep it free-flowing.
Conductivity cell-constant (K)	The product of the known electrical conductivity (EC) of a standard solution in a conductivity cell and the corresponding measured resistance (R) of the cell containing the standard solution. That is; $K = EC \times R$. The value of the cell constant is determined by the geometry of the cell and so is nearly independent of the temperature, but EC and R must be evaluated at the same temperature. Rearranging the equation and indicating temperatures by a subscript gives; $EC_t = K/R_t$.

Term	*Terminology*
Confounding	When the number of treatment combinations are large, the plot-to-plot variation within blocks on which is based the estimate of experimental error would thus increase and lower the precision of the experiment. The principle of confounding is to divide the replication in to two or more compact blocks of suitable sizes to ensure that the estimates of uncompounded effects are obtained with increased precision by removal of the effect of possible heterogeneity within a large replicate.
Conjunctive use of water	It refers to the use of saline water of wells in combination with canal water.
Conn	A nutrient storage enlargement at the base of the stem.
Conservation (resources)	The planning and management of resources in way so as to secure their wider use and continuous supply, maintaining their quality, value and diversity.
Conservation cropping	A way of farming that aims to maximise the protection against erosion that can be achieved through soil and crop management for sustained farm productivity.
Conservation tillage	Tillage designed to maintain roughness of a field surface and leave most of the previous crop residues on the surface while providing a suitable seed-bed and weed control for the next crop. The roughness reduces water run-off and water and wind erosion water exclusive of precipitation, stored soil moisture or ground water, which is required to meet evapo-transpiration demand of the crop.
Consumptive use	Quantity of surface and ground water absorbed by crops and transpired or used directly in the building of plant tissues, together with that evaporated from the cropped area, expressed in units of volume per unit area.
Contact herbicide	A chemical that kills primarily by contact with plant tissue rather than as a result of translocation.
	The herbicide that kills primarily by contact with plant tissue rather than as a result of translocation.
Contact placement (fertilizer)	The drilling of seed and fertilizer together while sowing.

Term	Terminology
Contingency cropping	Contingency cropping is growing of crops in aberrant situations like drought and floods. It aims at partial mitigation of misery by producing some food, feed and fodder to encounter emergency conditions. Growing of crops one after the other without seasonal fallowing.
Continuous grazing	The grazing of a specific range or pasture by livestock throughout a year or grazing season. The term is not necessarily synonymous with year-long grazing.
Contour	An imaginary line on the surface of the earth connecting points of the same elevation.
Contour border irrigation	Border strip irrigation practiced along the contour.
Contour bunding	The construction of small bund across slope of the land on a contour, so that the long slope is cut into a series of small ones and each contour bund acts as a barrier to the flow of water.
Contour cropping	The cultivation of crops along the contour of a slope.
Contour farming	A method of cultivation wherein operations including sowing are carried-out along the contour. It reduces run-off, conserves more moisture and increases crop yield.
Contour interval	The differences in elevation between adjustment contours on a map.
Contour irrigation	A method of irrigation where the water is applied to a field or orchard which has been divided into strips along the contours.
Contour planting	Planting row crops along the contours. Contour ploughing: Ploughing done along the contours. Contour strip cropping: The cultivation practice involving growing of a soil exposing and erosion permitting crop in strips of suitable widths alternating with strips of soil-protecting and erosion-resisting crops, along the contours.

Term	Terminology
Contour terrace	A low terrace formed along the contours on steep mountain slopes with inadequate natural cover to prevent erosion until natural cover becomes established.
Control	The process of limiting an infestation.
Control plot	A standard check plot used for comparing the relative performance of treatments under study.
Convection	Motions in a fluid resulting in the transport and mixing of energy and other properties of the fluid. In meteorology, convection usually refers to predominantly vertical motions of a fluid arising because of buoyancy differences.
Convectional rainfall	Rainfall caused by vertical ascent of air owing to intense heating of the earth's surface.
Conventional tillage	The combined primary and secondary tillage operations normally performed in preparing a seed-bed.
Conveyance efficiency	Ratio between water received at the inlet to a block of fields and that released at the projects head works; fraction. Usually expressed in percentage.
Cool-season grass	Grass species adapted to rapid growth during the cool moist periods of the year usually dormant during hot weather or injured by it.
Co-operative better farming	A type of co-operative farming where the land is not pooled and the cultivation is carried on by each farmer separately. A member is free to form his own way of farming except in respect of the purpose for which he has joined the society, *e.g.,* for irrigation, purchase of seed or marketing of produce etc.
Co-operative collective farming	A type of co-operative farming wherein the land is owned by the society and cultivation is carried out jointly The members work on the land under the direction of a managing committee. The profits are paid to the members in proportion to the work and capital contributed by each member. The right or share of individual member in the land is not recognized.

Term	Terminology
Co-operative farming	Co-operative farming means a system under which all agricultural operations or part of them are carried out jointly by the farmers on a voluntary basis; each farmer retaining right in his own land. The farmers pool their land, labour and capital. The land is treated as one unit and cultivated jointly under the direction of an elected management. A part of profit is distributed in proportion to the land contributed by each farmer and the rest of the profit is distributed in proportion to the wages earned by each farmer.
Co-operative joint farming	In this type of farming, the land of members is pooled for joint cultivation. The ownership of each member over his own land is recognized by payment of a dividend in proportion to the value of his land. The members work under the direction of the managing committee and each member receives for his daily labour.
Co-operative tenant farming	A type of co-operative farming wherein the land is held by the society and not by the members independently. The land is then divided into plots which are leased out to members. The society arranges for agricultural requirements, *e.g.,* credit, seeds, manures, marketing of the produce etc. Each member is responsible to the society for payment of the rent of his joint plots. He is at liberty to dispose off his produce in such a manner as he likes.
Coppice farming	The word coppice has been defined as the ability to regenerate by shoot or root suckers, or a forest so established. Thus coppice farming is a practice of intensive exploitation of land capability by regular harvesting of such trees which produce numerous shoots from the stump after cutting. Coppice sprouts usually grow vigorously because they are served by roots big enough to feed the former tree.
Correction strip	An irregular strip or area of land lying between contour strips.
Correlation	A relation between two variable quantities such that an increase or decrease of one is associated (in general) with an increase or decrease in the other.

Term	Terminology
Correlation coefficient	The degree of correlation ranges from +1 to –1. A correlation coefficient of 0 means that the two variables are not interrelated. An 'r' value of +1 or –1 indicates complete association. A positive (plus) correlation means that high values of one variable are associated with high values of the other. A negative (minus) correlation means that as one variable increases the other tends to decrease.
Cortex	The tissue region between the epidermis and vascular system of a plant.
Cosmic ray	Radiation originating outside the earth (including its atmosphere).
Cost of cultivation	Total expenditure involved in raising a crop, including rental value of the land.
Cotton bale	A pressed bale in ginning and pressing factory which weighs about 170 kg of lint.
Cotton boll	The fruit of cotton plant.
Cotyledon	The first leaves of the embryo, one in the monocotyledonous and two in dicotyledonous plant.
	The first leaf or pair of leaves (depending on whether the plant is monocotyledonous or dicotyledonous) of the embryo of seed plants. (Also referred as Cotyledon Leaves).
Coumarin	A chemical growth-inhibitor which has germination inhibiting capability.
Count	A count is the number of hanks of 764 m each, which weigh 0.45 kg. The fineness of yarn is usually expressed in terms of counts.
Count rate (nuclear)	The number of counts registered per unit of time.
Counter (nuclear)	A device for indication of, and often for recording nuclear radiation.
Counter radiation or back	A large portion of the radiation absorbed by the atmosphere radiation is reradiated back to the surface known as back radiation.
Counting efficiency (nuclear)	The fraction of disintegrations counted expressed in percentage, it is equal to the number of counts registered by the counter per 100 disintegrations in the source.

Term	Terminology
Counting rate meter (nuclear)	An instrument which indicates the average time rate of occurrence of events. Connected to a geiger tube, it will register counts/min.
Covariance	The mean of the product of corresponding deviation of two variates from their individual means. Estimated as the ratio of the sum of cross products to the corresponding number of degrees of freedom.
Cover crops	Crops which are grown primarily to cover the soil and to reduce the loss of moisture due to leaching and erosion by wind and water.
Critical difference	The square root of the error mean square, measures the standard error per plot due to uncontrolled environmental effects. The standard error of mean is determined by dividing standard error by number of observations entered into in calculating the mean. The standard error of difference when multiplied further by $\sqrt{2}$ and t-value (at required level of significance) at the error degrees of freedom is known as critical difference.
Critical period (water)	It is the period during which crop is affected severely due to moisture stress and the loss cannot be compensated by adequate moisture supply in other periods or stages.
Critical period of weed-crop competition	Critical period of weed growth can be defined as that shortest time span in the ontogeny of crop growth, when weeding will result in highest economic returns.
Critical period or critical (crop)	The period or the stage of development in the life-cycle of a stages crop or a plant when it is most sensitive to the deficiency of a production factor and is most responsive to the correction of the deficiency.
Critical photoperiod	It is the day-length value separating the inductive range of photoperiod from non-inductive range, for example, the critical photoperiod for flowering in Xanthium is 15 hrs.
Critical region	That portion of the area under curve which includes those values of a statistic which lead to rejection of the null hypothesis.

Term	Terminology
Critical value	A value of a diagnostic factor (q.v.) which forms the boundary between suitability rating classes.
Crop	Refers to plants sown and harvested by man for economic purposes.
Crop cafeteria	Crop cafeteria is the demonstration of identified efficient crops or varieties in an agrometeorological region or zone offering an opportunity to the farmer to choose a suitable crop or crop combination commensurating with the available resources and requirements.
Crop calendar	A list of the standard crops of a region in the form of a calendar giving the dates of sowing and various operations including harvesting to be done during the crop seasons in years of normal weather.
Crop canopy	It is the structure of aerial vegetative parts with special reference to the size, orientation, depth, density and arrangement of leaves influencing the penetration and interception of radiant energy.
Crop coefficient	Ratio between crop evapo-transpiration to pan-evaporation when crop is raised in large field under optimum growing conditions.
Crop competition	Two plants are in competition with each other when the grown of either one or both of them is reduced or their form modified as compared with their growth or form in isolation.
Crop duration	Days taken by a crop from germination to maturity. Crop ecology: The branch of 'plant ecology' which deals specifically with the study of the interrelation amongst crop plants and environment including management practices.
Crop ecosystems	Cropping systems ranging between mono culture and multi-species culture of field and garden crops, single or in combination and their relation to environmental conditions and management practices.

Term	Terminology
Crop environment	A crop environment may be regarded as having two components, the gross environment, which takes into account all environmental factors affecting crop growth and the current environment, which takes into account the general soil and atmospheric conditions outside the crop, and also the changes of those general conditions caused by the plant community.
Crop equivalent yield	It is the conversion of crop yields into one form to compare the crops grown mixed or intercropped or sequentially cropped. Conversion is done into monetary value or protein or carbohydrate or intrinsic energy value.
Crop evapo-transpiration	Rate of evapo-transpiration of a disease-free crop growing on a large field (one or more ha) under optimal soil conditions, including sufficient water and fertilizer and achieving full production potential of that crop under the given growing environment, includes water loss through transpiration by the vegetation, and evaporation from the soil surface and wet leaves, mm/day.
Crop intensification	The concept, approach method and process of growing more than one crop per year in the same field by increasing cropping intensity.
Crop log	A graphic record of the progress of a crop contains a series of physical and chemical measurements.
Crop logging	Foliar diagnosis comparing the nutrients status of comparable leaves of high and low-yielding crop plants (generally used in sugarcane in Hawaii).
Crop modeling and simulation	It is a simplified representation of a more complex reality in a crop-production system. It describes how inputs or variables enter, move, interact and are controlled within the system and how the output of the system is generated and affected by other input and controlled variables. 'Simulations' is the operation or manipulation of a model of system to estimate the consequences of selected strategies or combinations of input variables. If appropriate prices and costs are included, the economic consequences of selected alternatives can be studied.

Term	Terminology
Crop physiology	The branch of plant physiology which provides an understanding of the dynamics of yield development in crops in terms of growth analysis, plant development and yield characters in crop stands instead of in isolated plants.
Crop production	Crop production is concerned with the exploitation of plant morphological (or structural) and plant physiological (or functional) responses within a soil and atmospheric environment to produce a high yield per unit area of land.
Crop productivity	Economic yield or production of plant product of economic importance, expressed in standard units per unit area.
Crop residue	The portion of a plant or crop left in the field after harvest, or that part of the crop that is not used domestically or sold commercially.
Crop rotation (sequential cropping)	The growing of different crops on a piece of land in a pre-planned succession.
Crop season	The most favourable weather condition to get better yield.
Crop system	Comprises all components required for the production of a particular crop and the interrelationship between them and environment.
Crop technology	The theoretical and practical approach to crop production in which concepts of applied science and advanced techniques are related to the financial consequences of their use.
Crop water-use efficiency	The ratio of crop yield to the amount of water depleted by the crop plants through evapo transpiration.
Crop-efficiency zone	A geographical area characterized by high spread of a crop combined with high yields consistent with minimum variation from year to year.
Crop-growth rate (CGR)	It represents dry weight gained by a unit area of crop in a unit time expressed as $g\ day^{-1}\ m^{-2}$.
Cropping index	The number of crops grown per annum on a given area of land multiplied by hundred.

Term	Terminology
Cropping intensity	It is the ratio between total cropped area and actual net cultivate area expressed in percentage.
Cropping pattern	The yearly sequence and spatial arrangement of crops or of crops and fallow on a given area.
Cropping pattern design	The crop configuration or sequencing done on paper for year-round land utilization at a given area considering physical, biological, and socio-economic factors prevailing at that area.
Cropping scheme	The plan according to which crops are raised on individual plots of a farm with an object of getting the maximum returns from each crop without impairing fertility of the soil.
Cropping system	Pattern of crops taken up for a given piece of land, or order in which the crops are cultivated on piece of land over a fixed period and their interaction with farm resources and other farm enterprises and available technology which determine their makeup.
	The cropping patterns used on a farm and their interaction with farm resources, other farm enterprises and available technology which determine their makeup.
Cropping systems research	The research activities, mainly in farmers fields, that focus on the understanding of farmers' existing cropping systems; design, testing, and development of new improved cropping pattern and component technologies for selected environments to efficiently utilize available farm resources.
Cropping-pattern	Yearly sequence and spatial arrangements of crops and fallow, on a given area.
Cropping-pattern testing	The growing of a designed cropping pattern at a given site and evaluating biological stability, agronomic productivity, and economic profitability.
Cropping-pattern zone	It is developed to divide the country into homogeneous units using the entities like soil and climate besides physical and agronomic criteria subdivided on the basis of isothermic lines.

Term	Terminology
Crop-production strategy	It refers to a subset of the farm-production strategy that involves only crops as it components. It is described by the spatial and temporal arrangement of crops to be raised in a parcel of land (cropping pattern) and the procedure for growing each component crop. A crop production strategy can be either one of mono-cropping or multiple cropping.
Crop-residue management	Use of the non-commercial portion of the plant or crop for protection or improvement of the soil.
Crop-yield index	A measure of comparison of the yields of all crops on a given farm with the average yield of these crops in the locality. The relationship is expressed in percentage.
Cross Resistance	In the context of weeds, it is an expression of mechanism that confers plants the ability to resist or withstand herbicides from different classes.
Cross-drill (cross sowing)	To drill the seed in two directions usually at right angles to each other.
Cross-pollinate	To apply pollen of one flower to the stigma of another, commonly refers as pollinating the flowers of one plant by pollen from another plant.
Cross-resistance	Resistance to two or more herbicides or chemicals due to a single resistance mechanism is referred as cross resistance.
Crown	The point where stem and root joinin a seed plant.
Crubbing	The process of removing stumps, sunken stones, boulders and all roots over a minimum size and up to a specified depth.
Crude fiber	This term includes all the insoluble forms of carbohydrates found in feeding stuffs. It is made up of cellulose, lignin and some pentosans and represents the tough fibrous parts of plants.
Crude protein	All nitrogenous substances contained in feedstuffs which include the true proteins (composed of a number of amino acids) and non-protein nitrogen (NPN) compounds, such as amides.
Crumb	A soft, porous, more or less rounded natural unit of soil structure from 1 to 5 mm in diameter.

Term	Terminology
Crushing strength	Force required crushing a body of dry soil or dry aggregate.
Crust	A hard or brittle layer formed on the surface of many soils when dry.
Cryophyte	A plant which grows on ice or snow.
Crystal	A homogeneous substance bounded by plane surfaces having definite angles with each other and having a regular geometrical form with a definite chemical composition.
Cubing	Process of forming hay into high density cubes to facilitate transportation, storage and feeding.
Culm	Joined stem made up of nodes and internodes in alternate order.
	The jointed stem of a grass which is usually hollow except at the nodes or joints.
Cultivable or culturable command area	Gross command area minus uncultivated area. Command area is that which falls within irrigable area of a project.
Cultivar	It is equivalent to 'variety' and is defined as an assemblage of cultivated plants which is clearly distinguishable by any character (morphological, physiological, cytological, chemical or others) and when reproduced (sexually or asexually) retains its distinguishing characters. The term is derived from cultivated variety.
Cultivated land-utilization index	This is calculated by summing the products of land land area planted to each crop, multiplied by the actual duration of that crop and divided by the total cultivated land area available during 365 days.
Cultivation	It includes any labour and care taken in the raising of plants, such as stirring soil, fertilizing etc. or to loosen the soil around a plant for the primary purpose of weed control.
Cultivation factor	The number of years under cultivation as a percentage of the total cultivation/non-cultivation cycle. Expressed as R, in per cent.

Term	Terminology
Cultivation index (CI)	Ratio of organic carbon in the cultivated soil to that in virgin soil.
Cultural control (pests)	Control of pests by cultural practices such as soil tillage, management of fertility, weeds water and crop residue, plant spacing and cropping pattern.
Cultural practices	Crop-care practices including land preparation, seeding or transplanting, weed control, fertilizer and insecticide application, water control in the field etc.
Cumulative run-off	The total volume of run-off over a specified period of time.
Cup anemometer	It is a device for measuring the wind speed from the speed of rotation of wind mill which consists 3 or 4 cups with hemispherical or conical shapes each fixed to the ends of a horizontal arm projecting from a vertical axis.
Curing	Natural or artificial aging of the agricultural produce brought about usually by some (dry or wet) heat treatment.
Cusec	The quantity of water flowing at the rate of 1 cubic foot per second. One cubic foot of water weights 62.4 lbs or 6.24 gallons.
Cut off time (irrigation)	In surface irrigation, the time at which the supply of water is cut off at the top end, leaving the water already supplied to flow down the slope.
Cuticle	The outer corky or waxy covering of the plant.
	A thin, continuous, non-cellular, lipoidal membrane (varnish-like layer) covering the entire plant surface (shoot), formed by oxidation of plant oils.
Cutin	The chief structural component of cuticle, composed of polymerized long-chain fatty acids and alcohols, protecting the photosynthetic mechanisms of plant cell.
Cyclic salt	Salt brought into the soil by winds off the sea or inland salt lakes.
Cyclone	A system of winds blowing around the centre of low barometric pressure. The flow round a cyclone is anticlock-wise is Northern hemisphere and vice versa in Southern hemisphere.

Term	Terminology
Cyclotrun (nuclear)	A circular magnetic resonant accelerator.
Cycocel (CCC)	2-chloroethyl trimethyl ammonium-chloride. A plant growth regulator.
Cytokinins	A term used for substances which promote cell-division and exert other growth regulatory functions *e.g.* zeatin (naturally occurring) and 6-benzyl amino-purine (synthetic).
Cytoplasmic male sterility	A type of male sterility conditioned by the cytoplasm rather than by nuclear genes and transmitted only through the female parent.
Dairy husbandry	The care, breeding, feeding and milking of dairy cattle and production and sale of milk.
Dam	A barrier constructed across a river course for varied purposes, such as creating a reservoir, creating a head which can be used for irrigation, generating power, etc.
Dapog rice nursery	Rice seedlings are raised on well-prepared soil (wet bed) either in field or in nursery boxes, covered under banana leaves or plastics with bamboo pegs. Dapog seedlings are ready for transplanting in 9-14 days after the seed is sown. It is labour and time-saving method.
Darcy's law	The law stating that the velocity of a fluid in permeable media is directly proportional to the hydraulic gradient. In this law, hydraulic conductivity (K) is taken as proportionality constant.
Data	Numbers or measurements which are collected as result of observations.
Daughter nucleus (nuclear)	The nucleus to which a given radioactive nucleus is transformed when it decays.
Day cusec	The quantity of water in cusecs flowing for 24 hours. Day-length: Day-length is the interval between the sunrise and sunset.
Day-neutral plants	The plants that are independent of day length and can bloom under conditions of either long or short days.
Dead furrow	An open trench left in between adjacent strips of land after completion of ploughing.

Term	Terminology
Dead storage	It is the capacity of a reservoir below the outlet level.
Dealkalization (solodization)	The leaching of sodium ions and salts from the soil profile.
Decalcification	Removal of calcium carbonate or calcium ions from the soil by leaching.
Decay (nuclear)	The radioactive disintegration of a nucleus. Deciduous: Plants or trees that shed their leaves at a particular season, usually winter.
Deciduous Plants	The plants which lose their leaves during the winter.
Declarable content	The content of a nutrient which, according to national legislation, may be given on a label or document associated with a fertilizer or soil conditioner.
Decomposition	The breakdown of minerals and organic materials.
Deep percolation loss	Water that percolates downwards through the soil beyond the crop root zone.
Deep placement	Inserting, drilling or placing the fertilizer or nutrients below the soil surface or furrow at a depth of 20 cm or more to supply nutrients before planting of crop.
Deep seepage	The portion of the run-off which escapes from a reservoir through the underlying earth or rock strata below any possible inter- cepting cut out constructed at the dam.
Deep soil	A soil having a solum depth of more than 100 cm.
Deep tillage or sub soiling	Refers to tillage that loosens the soil 40 cm deep or more with the purpose of conserving moisture in the monsoon season or for breaking hard pans.
Deep water rice	Rice grown with more than 51 cm to 5-6 m of standing water.
Deferred grazing	Withholding animals from a pasture beyond the normal beginning of the grazing season.
Deficiency disease	An abnormality caused by the absence or subnormal level of availability of an essential nutrient element.
Deflocculate	To separate or breakdown soil aggregates of clay into their individual particles.
Defoliant	A chemical that causes the foliage to abscise from a plant.

Term	Terminology
Defoliant or defoliator	A chemical preparation intended for causing leaves to drop from crop plants, such as cotton, soyabean or tomato, usually to facilitate harvest.
Defoliation	A reduction in the normal amount of foliage either due to insect, fungal attack or other injury or spraying of chemical may be partial or complete (applies to leaves only).
Degeneration	Loss of vitality or vigour of a biological system due to physiological disorder, disease or other unfavourable conditions.
Degradation (pesticides)	The breakdown or disintegration of the chemical compound.
Degree day	It is a measurement of the departure of the mean daily temperature above the minimum threshold temperature for a crop or plant.
Degree of freedom	The number of degrees of freedom is the number of independent comparisons that can be made within the body of data.
Dehiscent	Splitting open of a pod or fruit in a characteristic manner at maturity.
Dehulling	The removal of husk or hull (outer covering) from intact grains or seeds.
Dehydration	The rapid removal of moisture to a very low level by applying external heat.
Delinter	A machine used to remove residual lint (fuzz) from cotton seed.
Delta (river)	A deposit of sand or sediment, usually triangular, formed at the mouth of rivers.
Delta (water)	The total depth of water required by a crop.
Demonstration	Practically showing to the user the working of a particular practice or technology developed and established on a research farm.
Demonstration trial	A non-replicated trial used to demonstrate some established fact or principle.

Term	Terminology
Denitrification	The biochemical reduction of nitrate or nitrite to gaseous nitrogen in the soil either as molecular nitrogen or as an oxide of nitrogen usually carried out by denitrifying bacteria, which result in escape of nitrogen to the atmosphere.
Denitrifying bacteria	A group of soil bacterial which breaks down nitrates an aerobically to produce nitrogen.
Deoxyribose	A 5-earbon sugar which is present in deoxyribose nucleic acid (DNA).
Deoxyribose nucleic acid	The nucleic acid of the chromosomes, the main carrier of (DNA) genetic information.
Depletion	The continued withdrawal of water from a surface or ground water stream, reservoir, or basin at a rate greater than the rate of replenishment.
Depletion phase (irrigation)	Portion of the total irrigation time between inflow shut-off and beginning of recession at upper end of the field.
Depth of cut	The maximum depth of the penetration of the tool measured with reference to the initial soil surface.
Depth of irrigation	Water depth required to bring soil water content of root zone to field capacity.
Dermal toxicity	Measure for the toxicity of a pesticide that will cause toxic symptoms when absorbed through the skin.
Desalinization	Removal of salts from saline soil, usually by leaching.
Desert	Places where rainfall is often less than 250 mm per year with extreme temperature fluctuations. Most of the vegetation is restricted to living only during the very short rainy season.
Desert crust	A hard layer, containing calcium carbonate, gypsum or other binding materials, which is exposed at the surface in desert regions.
Desert soil	It often contains high percentage of soluble salts, high pH, poor in organic matter; water is scarce and in this area usually coarse millets are grown.

Term	Terminology
Desiccant	Preparation intended to artificially speed up the drying of crop plant, parts such as cotton leaves and potato vines.
	A chemical that promotes or accelerates dehydration of plant tissue, causing drying; it also lowers moisture content of seeds to facilitate seed harvest
Design	A method of arranging sample or experimental plots to minimize the effects of uncontrolled variations in fertility and other natural factors and to make possible estimate of such effects in relations to those due to variations in treatment.
Design factor	Ratio between canal capacity and maximum discharge in m^3/sec and the maximum daily supply requirements during the peak water use period in m^3/day.
Desilting	The removal of ex silt and mud from farm ponds and reservoirs and usually carried out when water levels are low.
Desorption	Movement of particles of chemical substances (*e.g.,* herbicides) from the soil surface into the soil solution; the reverse of adsorption.
De-suckering (tobacco)	Removal of the suckers.
Detasseling	Removal of young tassels (pollen-producing organ) in maize plants before pollens are released so as to render. the plants fully female, especially in hybrid maize seed production
Detergent	A cleaning substance such as soap. Detergents are surface active agents (surfactants), many having applications in pesticide formulation as emulsifiers and wetting agents.
Determinants of cropping pattern	Environmental factors that influence the performance of cropping pattern and are not readily modifiable by changes in cultural techniques of crop production.
Determinate growth	The flowering of plant species uniformly within certain time limits. The plant first completes its vegetative phase and then enters into reproductive phase, thereby making a clear-cut distinction between these two growth phases, *e.g.,* cereals and grasses.

Term	Terminology
Detritus	Dead organic matter, mainly of fallen leaves as leaf litter in forests. The microbes decomposing detritus are called detritivores.
Deuterium	An isotope of hydrogen having one proton and one neutron in the nucleus. It is often called as heavy hydrogen.
Dew point	The temperature at which air on being cooled becomes saturated with water vapour.
Dew, mist and fog	Dew is the water that condenses and deposits on cool· surfaces due to saturation of atmospheric vapour. Fog and mist are low-hanging clouds in which water droplets are so small that they do not settle on horizontal surface. A fog consists of very small (0.01—0.1 mm in diameter) water droplets far enough apart that they do not fuse, and are suspended in saturated air, so that they do not evaporate and are light enough to settle. When thin it is called mist.
Diagnosis	The art of distinguishing and determining disease with the help of case history, symptoms and laboratory techniques.
Dialysis	It is the separation of colloids in solution from other substances by selective diffusion through a semi-permeable membrane.
Diara lands	The lands which are situated in between the natural levees that get inundated for different length of periods and are periodically eroded and formed by meandering, braiding and coarse changing of rivers.
Diastase	A group of enzymes which digest starch into glucose.
Dibbler	A device used for dibbling.
Dibbling	A method of sowing by placing crop seeds in the holes manually by using a dibbler wherein specific spacing and number of plants are maintained between the rows and within the rows.
Dicalcium phosphate ($CaHPO_4$)	A product containing not less than 4% of P_2O_5 in citrate-soluble form, which is considered available to plants.

Term	Terminology
Dicotyledon (Dicot)	Any seed plant having two cotyledons.
	Plants which have two seed leaves or cotyledons, generally includes the broad-leaved plants.
Diffused radiation	Solar radiation reaching the earth surface after having been scattered from the direct solar beam by molecules or aerosols in the atmosphere.
Diffusion	The movement of molecules from regions of high to regions of low concentration.
Diffusion pressure deficit (DPD)	It is the difference in diffusion pressure of a solution and pure solvent at the same atmospheric pressure.
Digestible crude protein (DCP)	Is the most common way of expressing the protein value and requirements for the ruminants? DCP of feed stuff is obtained by multiplying the crude protein content by the digestibility coefficient of the protein in the fodder.
Digestible dry matter	Feed-faeces differences expressed in calories.
Digestible energy	Feed-faeces differences expressed in calories.
Digestible nutrients	Portion of nutrients consumed, digested and taken into the animal body.
Digestion	The conversion of complex foods, generally insoluble to simpler substances which are soluble in water.
Dihybrid	Hybrid that is result of crossing of homozygous parents that differ with respect to 2 loci.
Dike (levee)	An embankment to confine or control water, especially one built along the banks of a canal or a river to prevent overflow of lowlands.
Diluent	Any substance used to dilute or carry an active ingredient or formulation.
	Any liquid or solid material used to dilute or carry an active ingredient in the preparation of a formulation or application.
Dimorphic	The ability of a plant to generate different sized shoots or leaves.
Dioecious	Plants with separate male and female individuals.
Diploid	A plant with 2 sets of chromosomes.

Term	*Terminology*
Direct application	Precise application to a specified area or plant organ such as to a row, bed, leaves or stem of the plant.
Direct drilling	The sowing of seeds directly into a field with a drill without previous cultivation and usually follows the application of weed-control sprays.
Directed Application	An application of a chemical directed to a restricted area, such as a row or a bed, at base of plants, minimizing contact with the crop.
Directed spray	An application made to minimize the amount of herbicide applied to the crop. This is usually accomplished by setting nozzles low with spray patterns intersecting at the base of the plants just above the soil surface.
Disease endurance or tolerance	The ability of plants to tolerate the invasion of a pathogen without showing much damage and any symptom.
Disease resistance	The ability of plants to withstand, oppose or overcome the attack of pathogen. It is variable in amount ranging from zero to cent per cent and therefore; different terms as given below are used to denote the different degree of resistance. Immunity: complete, *i.e.,* cent per cent resistance; practical resistance: high or moderate resistance, susceptibility: slight or zero resistance.
Disinfectant	A substance which can kill or inactivate pathogenic micro-organisms in the environment or on the surface of plant parts before infection.
Disintegration (nuclear)	A spontaneous radioactive transformation of one species of nucleus into a nucleus of a different type, usually accomplished by the emission of radiation.
Dispersal	Movement of seeds and other plant propagules to short or long distance, away from the mother plant.
Dispersed soil	Soil in which the day readily forms a colloidal soil, usually with a low permeability, which upon drying shrinks, cracks and becomes hard and upon wetting slakes readily and is plastic.

Term	Terminology
Dispersing agent	A material that reduces the cohesive attraction between like particles. Dispersing and suspending agents are added during the preparation of wettable powders to facilitate wetting and suspension of the toxicants.
Dissolved bone	Ground bone or bone meal that has been treated with sulphuric acid. It is commonly known as 'bone superphosphate' and contains 1-2 % nitrogen and 16% total P_2O_5 of which about 8% is water-soluble.
Distribution efficiency	Ratio of water made directly available to the crop and that released at the inlet of a block of fields.
Diversified cropping	A cropping plan in which no single crop contributes 50% or more towards the total crop production or monetary income (comparable equivalents) annually.
Diversified farm	A farm on which no single product or source of income equals as much as 50% of the total receipt and on such farm the farmer depends on several sources of income.
Diversion boxes	A water control device used for diverting irrigation water in different directions. It is made up of metal sheet or cement concrete structure available in 2-way, 3-way or 4-way diversion structure.
Diversion canal	A canal to divert water from one point to another. In irrigation practice, it extends from the point of diversion at the main canal to the beginning of the distribution system.
Diversity index	It measures the multiplicity of farm products which are planted by computing the reciprocal of sum of squares of the share of gross revenue received from each individual farm enterprise in a single year.
Dockage	Percent impurity in a seed sample.
Dolomite	A natural mineral composed of calcium and magnesium carbonate, widely used as a liming material and as an ingredient In fertilizer mixtures. Dolomite· is used in fertilizers to render them non-acid forming. It also supplies available magnesium.
Domestication	The process of bringing a wild species under human management.

Term	Terminology
Donnon equilibrium	An electrochemical equilibrium, regulated by electrical and diffusion phenomena and can account for passive accumulation of ions against concentration gradient.
Dormancy	It is a physiological phenomenon due to which plants remain alive but do not germinate. It refers to resting stage which could be due to unfavourable environmental factors or biochemical and physical factors.
Dormancy	The condition in which seeds or other living plant organs are not dead but do not grow even under conditions (moisture, temperature and oxygen) favourable for plant growth; temporary suspension of visible plant growth.
Dormant	State of inhibited growth or resting conditions of seeds or other living plant organs due to internal causes.
Double cropping	Taking 2 crops a year in sequence from the same piece of land.
	Growing two crops a year in sequence.
Double cross	A hybrid between 2 single crosses involving 4 different inbred lines.
Drage	A force retarding the velocity of water or wind over the ground surface.
Drain	A buried pipe or other conduit (closed drain) or a ditch (open drain) for carrying off surplus surface or ground-water.
Drainage	The removal of excess surface or ground water from land by means of surface or subsurface drains.
Drainage basin	An area from which surface run-off is carried away by a single drainage system. Also called catchment area, watershed and drainage area.
Drainage coefficient	The rate expressed as depth, in cm of water drained off from a given area in 24 hrs is known as drainage coefficient.
Drainage requirement	The amount of water that drains must convey in a given time for satisfactory crop growth.

Term	Terminology
Drainage water	Also called gravitational water, free water, and excess water. The water that the soil is unable to hold against the force of gravity. Also water .from surface, ground or storm water flowing into a drain.
Dredge com	Mixture of cereals, such as oats and barley, sown together.
Drift	The movement of dust or spray passively with the wind.
Drill	An equipment for sowing seeds in line.
Drill row	A row of seeds or plants sown with a drill.
Drilling	The process of placing seeds in rows at uniform rate and at controlled depth with or without covering them with soil.
Drizzle	Rainfall in which water drops are very small ranging from 0.2 to 0.5 mm in diameter.
Drop structure	A device made up of wood, iron or cement concrete installed wherever drops in the field levels, field channels for providing safe and non-erosive flow of water.
Drosometer	Instrument for measuring dew. It is also termed as Oetalies dew gauge.
Drought	The moisture deficit which results when the amount of water available in the oil is not sufficient to meet the demands of potential evapo transpiration.
Drought avoidance	Permits a crop to grow for a longer time in a given moisture-stress conditions (may be due to deep root system/uses less water).
Drought resistance	Relative ability to maintain growth and yield with little or no major setback under moisture-stress conditions.
Drought year	A year in which rainfall of particular place is short by more than twice the standard deviation.
Dry farming or dryland farming	The practice of crop production entirely with rain water received during the crop season or on conserved soil moisture in low rainfall (< 800 mm) areas of arid and semi-arid climate and the crops may face mild to very severe oisture stress during their life-cycle.

Term	Terminology
Dry season	A period during which water deficiency occurs, as stored water is used for evapo-transpiration and actual evapotranspiration falls below potential evapo-transpiration.
Dry spell	A period of at least 15 consecutive days, during which not a single wetting ram has fallen.
Dry-bulb thermometer	It is one of the 2 thermometers of a psychrometer whose bulb is bare and which indicates the air temperature.
Dryer	A unit which provides the conditions for reducing moisture generally by forced ventilation with or without addition of heat.
Drying	The reduction of moisture in a product usually to some predetermined moisture content.
Dry-matter content	Plant substances less water; found by oven drying (usually at 70°C) a weighed sample to a constant weight.
Dune	A mound or ridge of loose sand piled up by the wind usually occurs in desert areas.
Dung	The semi-solid excrement of animals used as manure and soil conditioner.
Durability	Durability of an agricultural system is defined as the extent to which the ecological basis of the system allows it to operate for an indefinite period.
Duripan	A subsurface diagnostic horizon cemented by silicon. Durum or macaroni wheat. It is the most important tetraploid wheat species of the group *Triticum durum,* and is next in importance to the hexaploid bread wheat. It is preferred for the preparation of vermicelli or semia.
Dust (pesticide)	This usually contains only 2-10% of the insecticide, rest inert material as talc and is used in areas where it is difficult to obtain large volume of water for spraying.
Dust mulch	A layer of fine, deflocculated soil material at the surface formed by excessive tillage; believed to be effective in reducing evaporation losses of water from soil.

Term	Terminology
Duty (water)	It is the area irrigated by 1 cusec discharge of water during the crop period; it is equal to twice the base divided by delta.
Dwarfing gene	Gene responsible for dwarfing characters in plants, *viz.,* Dee-geo-woo-gen is rice-dwarfing gene and Norin-10 is the wheat dwarfing gene.
Earth moving	Tillage action and transport operations utilized to loosen load, carry and unload soil.
Earthing up	The process of putting the earth or soil just near the base of stems of certain crops like sugarcane, cassava, papaya, potato, banana etc., to give support to the plants.
Easter	Any of a class of organic compounds formed by the reaction of an acid with an alcohol.
Ecad	A form of a plant modified by its habitat; the modifications are somatic (induced by environment) and thus not heritable.
Ecesis	Adjustment of an organism to a new habitat where it can complete its life-cycle. Ecesis comprises of all the plant processes (germination, growth and reproduction) exhibited by migrating species from the time it enters a new area until it is thoroughly established.
Ecoclimate	Climate under crop canopy is called ecoclimate. Ecological amplitudes. The range of environmental factors over which an organism or a physiological process can function. It denotes the tolerance limits.
Eco-farming	It is the potential for introducing mutually reinforcing ecological approaches to food production. It aims at the maintenance of soil chemically, biologically and physically the way nature would do it left alone. Soil would then take proper care of plants growing on it. "Feed the soil, not the plant" is the watch word and slogan of ecological farming.
Ecological crop geography	A branch of crop ecology which deals with the broad spatial distribution of crop plants and the rationale of such distributions in terms of physical and socioeconomic environment influencing the production of crops.

Term	Terminology
Ecological efficiency	The percentage of energy transferred from one trophic level to another.
Ecological equivalence	The situation in which two or more species can replace one another in the same ecological niche because of the similarity in their ecological amplitude.
Ecological niche	The position or functional status of an organism within its community and ecosystem, resulting in organisms' structural adaptation, physiological response and specific behaviours.
Ecological succession	A natural process by which different communities colonise the same area over a period of time in a definite sequence.
Ecology	The study of the mutual relations between organisms and their environment.
Economic poison (pesticides)	As defined under the Federal Insecticide, Fungicide, and Rodenticide Act, economic poison "means any substance or mixture of substances intended for preventing, destroying, repelling or mitigating insects, rodents, nematode, fungi or weeds, or any other forms of life declared to be pests and any substance or mixtures of substances intended for use as a plant regulator, defoliant or desiccant. As so defined, economic poisons are now known generally as pesticides.
Economic viability	Economic viability of a pattern can be determined by a budget analysis. The analysis includes cost of labour and purchased inputs for all operations specified as well as yields obtained. The profitability and returns to resource products are received from two distinct zones in addition to species specific to that zone.
Economic yield	Economically useful part of total dry matter produced by a plant or crop in unit land area.
Ecophone	Population of individuals which although belong to the same genetic stock but differ markedly in vegetative characters, such as size, shape, leaves etc., due to the influence of environment.

Term	Terminology
Ecosystem	An ecosystem has been defined as an open system comprising plants, animal, organic residues, atmospheric, gases, water and minerals that are involved together in the flow of energy and circulation of matter. An ecosystem results from the integration of all living and non-living factors of the environment for a defined segment of space and time. It may also be defined as the whole complex of plants and animals forming a community together with all the physiological factors of environment which forms a single unit.
Ecotone	A transition zone between two adjacent biomes, containing some organisms from adjacent biomes and some characteristic ones restricted to the zone itself.

A narrow transitional zone of overlapping communities of the adjoining areas of distinctly different habitats or a zone where two or more diverse communities from different ecological zone co-exist. Ecotoners are normally rich in species as migrants are received from two distinct zones in addition to species specific to that zone. |
Ecotype	Plant type or strain within a species, resulting from exposure to a particular environment.
Edaphic	A term pertaining to the influence or relationship of soil or other similar media to plant growth in contrast to atmospheric influences.
Edaphic factors	Soil factors.
Edaphology	The scientific study of the relationship between soils and living things, including man's use of the land.
Edible oil	Any vegetable oil which is fit for human consumption.
Effective field capacity (tillage)	The actual area covered by the implement based on its total time consumed and its width.
Effective full ground cover	Percentage of ground cover by the crop when ET of crop is approaching maximum-generally 70 80% of surface area.

Term	Terminology
Effective half-life (nuclear)	The half-life of a radioisotope in a biological system as a result of the combined effects of the biological half-life and the radiological half-life.
Effective irrigation	A controlled and uniform application of water to crop land in required amount at required time, with minimum cost to produce optimum yields without waste of water and adverse effect on soil in form of soil salinity and water logging problems.
Effective outgoing radiation	The difference between the outgoing long-wave radiation from the earth's surface and downward long-wave radiation from the atmosphere.
Effective rainfall	A fraction of total precipitation which forms a part of the crop consumptive use or a fraction that becomes utilizable for crop production.
Effective rooting depth	Soil depth from which the full grown crop extracts most of the water needed for evapo transpiration.
Effective soil depth	Depth of soil up to which plant roots penetrate to draw water and plant nutrients.
Efficient cropping zone	It is a cropping zone in which relative yield index is 200% and relative spread indices vary from 90-200%.
Einstein	A unit for measuring the energy output of light that is equal to 1 mole of quanta.
Elasticity of production (Ep)	Elasticity of production is the percentage increase in output as compared to the percentage increase in input.
Electrical conductivity (EC)	The property of a substance to transfer an electrical charge (reciprocal of resistance). Used for the measurement of the salt content of an extract from a soil when saturated with water. Measured in millimmhosl cm at 25°C.
Electromagnetic radiation	Radiation consisting of electric and magnetic waves that travel at the speed of light.
Electromagnetic spectrum of light	Visible light which is composed of a wavelength spectrum for 390 nm to 760 nm with violet, indigo, blue, green, yellow, orange and red monochromatic light.

Term	Terminology
Electron	An elementary particle of nature having a unit negative charge and a mass of 9.1 x 10^{-28} g.
Electron volt	A unit of energy equal to the energy acquired by a singly charged particle when it is accelerated through a potential difference of one volt.
Electrophoresis (cataphoresis)	The movement of colloidal particles suspended in a fluid (dispersed phase) either towards the cathode or anode under the influence of an electric field.
Element	Substances which consist all matter at and above atomic level. An element consists of atoms of same atomic number.
Elevation	Vertical distance of a point or level on or affixed to the surface of the earth, measured from mean sea-level.
Elite variety	An improved variety developed by plant breeders and released to farmers due to its superiority in at least one respect.
Eluviation	Removal of soil material from the upper to the lower horizon in solution or in colloidal suspension.
Emasculation	In bisexual flowers, the removal of stamens before they burst and have shed their pollens. Its purpose is to check self-pollination and is done before crossing.
Embryo	The rudimentary plant within the seed.
Embryo culture	A technique of excising the embryo from the seed at some stage during its development and germinating it on a special medium.
Emergence	Coming out of a place, such as a germinating seedling breaking through the soil surface or a flower from a bud or the adult insect leaving the last nymphal skin or pupal case. Appearance of the first leaves of the crop plant coming through the ground.
Emergent plants	Type of plant species that grows above the water surface in water bodies, such as cattails and water lilies etc.

Term	Terminology
Emerson effect or photosynthetic enhancement effect	The effectiveness of long wave-lengths of light (+680 nm) can markedly increased by the simultaneous application of short wave-lengths of light (680 nm). The effect of the two superimposed beams of light on the rate of photo- synthesis exceeds the sum effects of both beams of light used separately.
Emmer wheat	Tetraploid wheat of the group *Triticum dicoccum* which is grown on a limited area in southern India and is preferred in the preparation of *suji* or *rawa.*
Emulsifiable concentrate	Produced by dissolving the toxicant and an emulsifying agent in an organic solvent.
	An emulsified concentrate is an oil-based liquid compound containing a high concentration of insecticide compound. It is a mixture of insecticides solvents and emulsifiers to make it easily mixed in water, and wetting and sticking agents to make the material cover and adhere to the plants.
Emulsifier	A substance, which promotes the suspension of one liquid in another.
	The material that facilitates suspension of one liquid in another.
Emulsifying agent	A substance which stabilizes (reduces the tendency to separate) a suspension of droplets of one liquid in another liquid which otherwise would not mix with the first one. Also referred to as an emulsion stabilizer or emulsifier.
Emulsifying Concentrate (Ec)	A single phase liquid formulation that forms an emulsion when added to water.
Emulsion	A dispersion of fine particles of oily material in water (OW type) or less commonly in pesticide usage a dispersion of water droplets in oil (WO type).
	A mixture in which one liquid is suspended in minute globules in another liquid without either losing its identity (*e.g.,* oil in water).
Encapsulated formulation	Herbicides enclosed in capsules of thin polyvinyl or other material to control the rate release of chemical and thereby extend the period of activity.

Term	Terminology
Endemic (a plant or animal)	Confined to a particular area either because it is evolved there or has not been able to spread more Widely or because it had survived there and nowhere else.
Endodynamorphic	Soils whose properties are influenced mainly by parent material.
Endoparasite	A parasite which enters the host and feeds from within the host.
Endosperm	The main portion of the grain kernel with the exception of the small space occupied by the germ, the endosperm tissue, fills the interior of the kernel. It is resulted from triple fusion and is the nutritive material on which the embryo feeds on, when germinating.
Endotrophic (mycorrhiza)	An association between fungi and plants wherein fungi live intracellulariy for most part of their life.
Energy	The capacity to perform work. It exists in many different inter-convertible forms such as heat, light, motion and electricity.
Energy balance	It is the balance between the net radiation and soil heat flux, sensible heat and latent heat.
Energy cropping plantation	This includes growing of crops and trees for energy purpose, *e.g.,* Leucanea, Casuarina, Eucalyptus, etc.
Energy crops	To obtain energy such as ethanol and alcohol.
Energy farming	A concept involving the farming of fast growing plants or trees for the purpose of providing biomass that can be used directly as fuel or converted into other forms of fuel or energy products.
Energy productivity ratio	Ratios show food yield per unit of energy input and is most often used in evaluating the energetics of agricultural systems. The energy productivity ratio is food production in nutritional equivalents to fossil-fuel-based feedback.
Energy-efficiency index	It measures the rate at which the use of energy inputs generates energy outputs. It is comparable to the economic efficiency index.

Term	Terminology
Energy-rich phosphate bond	A bond joining a phosphate group to an organic molecule; when the bond is broken, energy is released.
Enriched compost	Compost fortified with the addition of fertilizers (urea, SSP, rock-P) during its production to raise its nutrient content and narrow down the C:N ratio.
Enrichment (soil)	General term for addition of nutrient material to a soil body.
Enrichment or atom percent excess (nuclear)	The abundance of a given stable isotope in a labelled sample minus the abundance of that isotope in nature.
Enterprise (farm)	It speaks of any farm activity tor which a separate economic analysis is feasible and meaningful, *e.g.,* a single crop, a crop mixture, the raising of animals, etc.
Entisols	Solid that has no diagnostic pedogenic horizons. They may be found in virtually any climate on very recent geomorphic surfaces.
Environment	The sum total of all the external forces or factors, both biotic and abiotic that affects the physiological behaviour or performance of an organism or of a group of organisms.
Environment index	This is calculated by subtracting the mean yield of all locations from the treatment mean yield of any one location; a positive value indicates that the location is better than average and a negative index the poorer than average.
Environmental complex	A union of sites when cropping pattern determinants is the same.
Environmental factors	Factors over which farmers have little direct control, including the physical, biological, and socio-economic aspects of their setting.
Enzyme	A complex organic substance, a protein, produced by cells and having the power to initiate or accelerate specific chemical reactions in the metabolism of plants arid animals.
Ephemerals	Many desert plants which germinate at the beginning of the rainy season and have an extremely short growing period called ephemerals.

Term	Terminology
Epicotyl	That portion of an embryo or seedling above the cotyledons or cotyledonary node, plumule.
Epigeal germination	The type of germination in some seeds such as beans, gourd, castor, cotton, etc., where the cotyledons are seen to be pushed above the surface by the rapid elongation of the hypocotyl.
Epinasty	The increased growth on the upper surface of a plant organ or part (especially leaves) causing it to bend downward.
	The twisting or curling of leaves and stems caused by uneven growth of cells, especially in leaves, in which the upper surface grows faster than the lower surface, thus causing the leaf edges to bend curl downwards.
Epiphyte	A plant which grows on another plant but does not secure nourishment from it.
Eppley pyreheliometer	An instrument for measuring both direct and diffuse radiation.
Equivalence factor	In intercropping, for component 'a', the equivalence factor is the number of plants of component 'a' which is equally competitive to one plant of component 'h'. If a given component has an equivalence factor less than one, it means that it is more competitive (on a plant-for-plant basis).
Equivalent acidity	The number of parts by weight of $CaCO_3$ required neutralizing the acidity resulting from the use of 100 parts by weight of fertilizer.
Equivalent basicity	The number of parts by weight of $CaCO_3$ that corresponds in acid neutralizing capacity to 100 parts by weight of the fertilizer.
Equivalent conductivity	The conductance of a solution which contains 1 g equivalent of the solute when placed between the two electrodes of definite size and 1 cm apart. It is represented by and is numerically equal to specific conductance K multiplied by volume in CC containing one gram equivalent of the solute.
Equivalent weight of a soil	The weight of a clay or organic soil colloid which has a combining power equivalent to 1 g-atomic weight of hydrogen.

Term	Terminology
Eradication (disease)	A method of disease control in which the pathogen is eliminated after it has established.
Erectophile	Steeply inclined leaves which intercept least light per unit of leaf-area index.
Erg	It is CGS unit of work and is defined as the amount of work done by a force of 1 dyne in moving the point of application through 1 centimeter (1g=981 dyne). Erg is a very small unit and hence a larger unit called Joule is commonly used, 1 Joule = 10^7 ergs.
Erodibility	The lack of resistance of soil to soil erosion.
Erosion	The wearing away of the land surface by water, wind or other forces.
Erosion accelerated	Abnormally rapid erosion in an environment disturbed by animal life, chiefly man.
Erosion rill	Water erosion producing very small and numerous channels.
Erosion sheet	Erosion of a fairly uniform layer of material from the land surface, often scarcely perceptible especially when caused by wind.
Erosion splash	A form of soil erosion resulting from soil splash under the impact of falling rain drops.
Erosion-resistant crop	A crop, because of its dense foliage, root system etc. provides effective protection against soil loss by erosion.
Erosivity	The ability of rainfall to cause soil erosion.
Erucic acid	(C is 13-Docosenoic acid) C_n fatty acid with a double bond; a homologue of oleic acid with 4 more carbons. Oils with a high erucic acid content are considered unsuitable· for human consumption.
Essential elements	Elements obtained by a plant from the soil and air and without which plant cannot complete its life-cycle.
Ester	Any of a class of organic compounds formed by the reaction on an acid with an alcohol.
Etiolation	Refers to the condition of a plant grown without light. The plant lacks chlorophyll, has an excessively elongated, weak stem and under-developed leaves.

Term	Terminology
Eutrophic	Soil solution (or waters) containing nearly optimal concentrations of nutrients.
Evaporation	It is a process involving conversion of liquid into vapour.
Evaporimeter	Device for measuring the amount of water evaporated into the atmosphere in a given time interval.
Evapo-transpiration	It is the total amount of water lost due to transpiration by a crop and evaporation from the soil surface during a specified time from a particular area.
Evolution	The development of a race, species or other group.
Exchange capacity	The total ionic charge of the adsorption complex active in the adsorption of ions.
Exchangeable cation (ECP)	This term indicates the degree of saturation of the soil or percentage exchange capacity with a cation.
Exchangeable phosphate	The phosphate anion reversibly attached on the surface of the solid phase (anion-exchange capacity) of soils, in such form that it may go into molecular solution by anionic equilibrium reactions with other anions of the solution phase.
Exchangeable potassium	The potassium which is held mainly by the colloidal portion of the soil and is easily exchanged with the cation of the neutral salt solutions less the water-soluble potassium. It is readily available to growing plant roots.
Exchangeable sodium (ESP)	The degree of saturation of the soil exchange complex with percentage sodium expressed in percentage.
Excitation (nuclear)	The transition of a nucleus, an atom or a molecule to an energy level above that of its ground state.
Excited state	When a molecule electron absorbs light quantum, the free energy increases. It is then at a higher energy level called excited state.
Excretion	The release of substances from the organ of a plant into its environment, for example, roots may excrete alkaloids, vitamins, nucleotides, amino acids, auxins, sugars, organic acids etc.

Term	Terminology
Exhaustive crops	Such crops leave the field exhaustive after growing.
Exotic (plant)	A newly introduced plant not native to a place.
Experiment	A systematic procedure for making observations under controlled conditions, in such a way that they can be used for arriving at general conclusions regarding the population under study.
Experimental design	A logical structure of an experiment, that helps in obtaining results with some precision.
Experimental yield	The yield level of a crop variety obtained at experimental stations where yield maximization is the major objective using all possible sophisticated technology.
Extension	Conveying to farmers the results of agricultural research and other methods of better farming through demonstration and other media and assisting them in the adoption of better practices.
Extrapolation area	Adaptation domain of a cropping pattern composed of land types to which the cropping pattern is adapted.
Extra-terrestrial radiation	Amount of solar radiation received on a horizontal place at the top of the atmosphere.
Extreme temperatures	Highest and lowest temperatures attained during a given time interval.
F_1	The first filial generation of a cross between two individuals.
Flotation	It is a process to remove light materials like bran from heavy materials like wheat by adding a flotation agent in the water.
Fluating plants	Type of plants that grow in water and remain in floating conditions, such as water-hyacinth and water lettuce etc.
F_2	The second filial generation produced by crossing or self fertilizing individuals of the F_2 generation.
Factorial design	An experimental design in which at least two series of treatments are used is being combined in every possible combination.

Term	Terminology
Facultative anaerobes	Micro flora that can grow either with molecular oxygen or anerobically when supplied with a suitable electron acceptor or other molecular oxygen.
Facultative weed	Weed of wild community origin, escaping sometimes to crop field.
Fallow	It is the practice of allowing crop land to lie idle during a growing season to build up the soil moisture and fertility contents, so that a better crop can be produced the following year. It is usually worked periodically to control weeds and improve moisture infiltration.
Fallow land	A land left unfarmed for one or more growing seasons to kill weeds, make the soil richer.
Fallow systems	It is the sequence of crop years with fallow years. Extensive fallow systems are shifting cultivation systems. Intensive fallow systems are bush fallow and grass fallow systems.
Family	In plant taxonomy it refers to a group of genera.
Farm	An area of land used for agriculture either to raise a crop or pasturage or maintain livestock.
Farm budgeting	A process of estimating costs, returns and net profit on a farm.
Farm enterprise	An individual crop or animal production function within a farming system which is the smallest unit for which resource use and cost return analysis is normally carried out.
	An individual crop or animal production function within a farming system which is the smallest unit for which resource use and cost return analysis is normally carried out.
Farm forestry	Farm forestry is a process in which trees are grown specifically for fuel, food and for a variety of other resources on a farm. It is a three dimensional farming and provides an alternative to monoculture grain farming. The three dimensions are defined as the trees themselves, the harvest from the trees which is used to feed livestock and the animals and their products.

Term	Terminology
Farm management	The branch of agricultural economics which deals with the business principles and practices of farming with an object of obtaining the maximum possible return from the farm as a unit under a sound farming programme.
Farm planning	A process involving many decisions to be taken in respect of the kinds of crops to grow, rotations, mixtures, soil and water conservation practices to be followed and building, bullocks, machinery purchase etc.
Farm pond	A small body of water retained behind a small dam or held in a pit dug in the ground by water-harvesting techniques.
Farm receipts	The total amount of money received from the sale of farm products, the increase in inventories of livestock, feed suppliers and receipt from outside labour, rent of building etc.
Farming system research	It is a highly location-specific research which is multi- and interdisciplinary in nature and uses whole farm approach for improved technologies to enhance and stabilize agricultural production. The research strategy includes base data analysis, on-centre research and on-farm research. This is the final evaluation of system in the real world situation of the farmer.
Farming systems	Farming systems represent an appropriate combination of farm enterprises, *viz.,* cropping systems, livestock, fisheries, forestry, poultry, and the means available to the farmer to raise them for profitability. It interacts adequately with the environment without dislocating the ecological and socio-economic balance on one hand, arid attempts to meet the national goals on the other.
Farming systems research (FSR)	The study of whole farm systems, which include all the enterprises on the farm, their biological, economic and cultural operations and usually imply some involvement of the farmer in the research process.

Term	Terminology
Farm-irrigation efficiency	It is the ratio expressed in percentage of irrigation water available for crop production to that delivered at the farm head gate.
Farmyard manure spreader	A machine to carry farmyard manure and spread it in a regulated quantity.
Fast neutrons (nuclear)	Neutrons having energies in excess of about 0.1 meV.
Fats	These are esters of fatty acids with glycerol (glycerides) that are solid at room temperature.
Fatty acids	Fatty acids are monobasic ads containing only the elements carbon, hydrogen and oxygen and consisting of an alkyl radical attached to the carboxyl. The saturated fatty acids have the generic formula $C_2H_znO_2$.
Fatty oil	The extactable non-volatile oil of the plant which comprises the fat (glyceride) constitution.
Fauna	All the animal life in a given region for a period of time.
Fecundity	The ability of flowers to produce seeds that will germinate.
Feeding value (FV)	Animal production = f (Intake x feed utilization).
Fencing	A device that fixes the boundary of the farm and helps to stop the encroachment of the land and protection from wild animals and cattle.
Fermentation	Breakdown of complex organic compounds into simple compounds by microbial fermentation.
Ferrugenous soil	Soil containing large amounts of iron minerals, especially limonite and hematite, characterized by red colour or shades or yellow and brown.
Fertigation	Application of fertilizers along with irrigation water.
Fertility (soil)	The ability of a soil to supply all the essential nutrients in optimum amount and balance and in a form readily available to plants.
Fertility gradient	The variation in natural fertility in any direction across an area of land.
Fertilization (plant)	The fusion of two gametes to form a zygote.

Term	Terminology
Fertilization (soil)	The application of fertilizers to the soil that aid in the nutrition of plants. A material in which declared nutrients are in the form of inorganic salts obtained by extraction and/or by physical and/or chemical industrial processes.
Fertilizer	Any organic or inorganic material which is added to a soil to supply certain elements essential to the crop growth.
Fertilizer broadcaster	A fertilizer distributor with a spreading width substantially greater than the width of the machine.
Fertilizer distributor	A machine which distributes fertilizer at regulated and selected rates.
Fertilizer drill	A machine to deposit fertilizer in soil at regulated and, selected rates and at predetermined depth.
Fertilizer grade	An expression showing the legal guarantee of its available plant nutrients expressed as percentage of plant nutrients in a fertilizer, *e.g.,* a 10-5-5 grade of fertilizer indicates 10% nitrogen, 5% phosphoric acid (P_2O_5) and 5% potash (K_2O).
Fertilizer gun	A machine to throw a jet of fertilizer by mechanical, pneumatic or other means usually fitted to the side of the machine.
Fertilizer legislation	This refers to laws and regulations enforced by the Government to regulate the quality of fertilizers or fertilizer mixtures sold to consumers. According to the Fertilizer Control Order of the Government of India of 1957, each brand and grade of commercial fertilizer and fertilizer mixture is to be registered and each bag of fertilizer to be labeled with fertilizer grade clearly written on it.
Fertilizer mixture	A product obtained by physically mixing different fertilizers so as to contain more than one of the three major nutrients, *viz.,* nitrogen, phosphorus and potassium.
Fertilizer ratio	The fertilizer ratio designates the relative proportion of three major plant nutrients, keeping the percentage of nitrogen as one in the ratio. Thus, a 5-10-5 fertilizer mixture which contains 5% nitrogen, 10% P_2O_5 and 5% K_2O, has a nutrient ratio 1:2:1.

Term	Terminology
Fertilizer requirement	The quantity of certain plant nutrient elements needed in addition to the amount supplied by the soil, to increase plant growth or crop yield to a designated optimum.
Fetch (also generating area)	The length of fetch area, measured in the direction of the wind from the site in question, which is required to eliminate the effect of advection etc.
Fiber	Any thread like material, specially those thread like tissues having sufficient toughness for the use in the textile or similar industries. They may be derived from the vegetable, animal or mineral kingdom.
Fibrous root system	The root system that is composed of profusely branched roots with many lateral rootlets but with no main or tap root development.
Fibrous root-system	A root-system growing from base of a stem in which all roots have about the same diameter which are mostly slender, thread like as in grasses.
Field capacity	It is the moisture content in percentage of a soil on oven dry basis when it has been completely saturated and downward movement of excess water has practically ceased, which usually takes place within 2-3 days after saturation. It is the upper limit of the soil moisture available to the plant growth.
Field crops	Herbaceous plants grown on cultivated fields or crop plants that are grown on a vast scale.
Field efficiency	It is the ratio of effective field capacity and theoretical field capacity expressed in percentage.
Field irrigation requirement of crop	The per cent moisture content in a field sample of soil at any one time.
Field moisture deficiency	The quantity of water which would be required to restore the soil-moisture content to field moisture capacity.
Field strip cropping	A system of strip cropping in which crops are grown in parallel strips laid out across the general slope but which do not follow the contour. Strips of grass or close growing troops are alternated with field crops, often.

Term	Terminology
Field systems	It is practiced where one arable crop follows another and where established fields are clearly separated from each other.
Field water balance	Sum of all gains and losses of water over a given period of time, mm/period.
Field water capacity	The maximum amount of capillary water that a particular soil is able to hold.
Field-application efficiency	Ratio of water made directly available to the crop and that received at the field inlet.
Field-channel efficiency	Ratio between water received at the field inlet and that at the inlet of a block of fields; fraction.
Field-irrigation efficiency	It is the ratio expressed in the percentage of irrigation water available for use of crops to that delivered to a field.
Fifteen atmosphere percentage	Is the moisture percentage on dry weight basis of a soil sample which has been wetted and brought to an equilibrium in a pressure membrane apparatus at 15 atm pressure (221 lb/sq. inch). This characteristic moisture value for soils approximates the lower limit of water available for crop growth (permanent wilting percentage).
Fifty per cent yield decrement value	The measured value of the soil salinity or alkali that decreases crop yield by 50 per cent as compared with yields of the same crop on non-saline and non-alkali soils under similar conditions.
Filler	Any material mixed with fertilizers for any purpose other than the addition of available nutrients, such as for conditioning to give anti-caking properties and for increasing the weight to bring the percentage of nutrients to desired values.
Film badge (nuclear)	Small photographic film in light, tight envelope worn by personnel to register exposure to ionizing radiation.
Filter strip	A strip of permanent vegetation of sufficient width and vegetation density above farm ponds, diversion terraces and other structures to retard the flow of runoff water, causing it to deposit soil, thereby preventing silting of structure or reservoir below.

Term	Terminology
Final sample, laboratory sample	A representative part of the reduced sample or, where no intermediate reduction is required, of the aggregate sample.
Fine texture	Consisting of or containing large quantities of the fine soil fractions, particularly silt and clay. Includes sandy clay, silty clay, and clay textural classes.
Fire curing	A process of curing tobacco, specially for tobacco, used for chewing purposes. In this process leaves are wilted for a few hours in the field tied into bundles and hung on rod in a smoke hut then smoked for 12 hr by burning.
Firing	A process of applying heat to dry a substance rapidly as in processing tea.
Fishmeal	Commercial feed prepared by crushing clean, dried, undecomposed whole fish or fish cutting with or without the oil extracted.
Fission (biology)	A simple form of a sexual production occurring in unicellular organism whereby one cell divides to form two cells.
Fission (nuclear)	Splitting of a heavy nucleus into two lighter nuclei of about equal size, whereby one or two neutrons and a relatively large amount of energy (as well as some gamma radiation) are released.
Fixation (nutrient)	It is a process in a soil by which certain chemical elements essential for plant growth are converted from a soluble or exchangeable form to a much less soluble or to a non-exchangeable form, for example, phosphate fixation.
Fixation of nitrogen	The fixation of atmospheric nitrogen in the soil by symbiotic and non-symbiotic bacteria.
Fixation of phosphorus	The conversion of soluble phosphorus nutrient in the soil into less soluble and unavailable forms.
Fixed phosphorus	(1) That phosphorus which has changed to less soluble forms as a result of reaction with the soil; moderately available phosphorus. (2) Applied phosphorus, form not taken up by plants during the first cropping year, (3) Soluble, P attached to the solid phase of soil in forms highly available to crops; (4) unavailable phosphorus, P in other than readily or moderately available forms.

Term	Terminology
Fixed potassium	Potassium held by a soil which is neither water soluble nor readily exchangeable.
Flag stage	It may be defined as: (i) the early post-emergence stage of onion seedlings between the 'crook' stage and the emergence of the first true leaf. The bent tip of the seed leaf resembles a flag attached to a staff. Also referred to as the 'knee' stage. (ii) in cereals it refers to the stage when the upper most leaf is out but the inflorescence is still in the boot.
Flaming	Killing green plants by momentary exposure, to very high temperature from a flame thrower.
Floating rice	Tall rice grown where maximum water depth ranges between 1 and 6 m or more than half of the growth duration.
Flocculate	To aggregate individual particles into small groups or granules.
Flood irrigation	It is a method of irrigation wherein entire land surface is flooded.
Flopping	It refers to wilting of tobacco due to waterlogging.
Flora	The plants or micro-organisms of a particular area or a descriptive list of plants in an area including key for identification.
Flowering (anthesis)	Begins with protrusion of the first dehiscing anthers in the terminal spikelets.
Flue curing (tobacco)	The process consists of three distinct steps, viz., yellowing, fixing the colour and drying of the leaves.
Flux	The amount of an entity received by or transferred across a particular plane per unit time.
Fluxial rice lands (low-land rainfed swamps)	These are located in lower aspects of the land scape or in flat areas and are flooded during the greater part of the growing season.
Flyash	Finely divided particles of ash in fuel gases resulting from combustion of fuel or other material.
Fodder	Maize, sorghum or other coarse grasses harvested with seed and leaves and cured for animal feeding.

Term	Terminology
Fodder crops	The cultivated plant species that are utilized as livestock feed in the form of silage and hay.
Foliage	Also called leafage. The leaves of a plant are collectively called foliage.
Foliar diagnosis	An estimation of mineral nutrient deficiencies or excesses in plants based on examination of the chemical composition of selected plant parts, and the colour and growth characteristics of the foliage of the plants.
Foliar fertilization	Fertilization of plants, or feeding nutrients to plants, by applying chemical fertilizers to the foliage usually in the form of spray.
Foliar spray	Any spray applied to plants when leaves are present.
Food	An organic substance which contributes materially to growth and repair of tissues and which yields energy when respired.
Food chain	A series of plant or animal species in a community each of which is related to the next as a source of food.
Food poisoning	A general term applied to all stomach or intestinal disturbances due to food contamination with certain micro-organisms or their toxins.
Food reserves or reserve carbohydrates or non-structural carbohydrates or accumulate carbohydrate	Perennial and biennial legumes and grasses store energy as readily available carbohydrates in various plant parts. The reserve carbohydrates are primarily used to initiate regeneration after cutting, defoliation and dormancy. They are also used to develop heat and cold resistance and to promote flower and seed formation. Accumulated carbohydrates imply that they may or may not be the causative factors in regrowth.
Food web	Interlooking pattern of several interlinked food chains.
Foot-candle	This is the illuminance provided by a light source of the standard candle at a distance of one foot one ft candle = 10.764 lux.
Forage crop	Crop grown primarily for livestock feed, to be either harvested for hay, silage or green feed or harvested by grazing animals.

Term	Terminology
Forage forestry	The concept of forage forestry can simply be defined as: using the same land or a portion of it, simultaneously or sequentially for fodder, food and fuel.
Forage quality	Forage quality is defined as output per animal and is a function of voluntary intake and digestibility of nutrients when forage is fed alone and ad libitum to a specified animal. The terms like rate of consumption of digestible dry matter or digestible energy consumption have come into common usage these days to express forage quality.
Forage-crop modeling and simulation	A model is a simplified representation of a more complex reality in a forage-livestock production system. It describes how input or variables enter, move, interact and are controlled within the system and how the output of the system is generated and affected by other input and control variables. Simulation is the operation or manipulation of a model of a system to estimate the consequences of selected strategies or combinations of input variables.
Foreign body	Foreign objects including pieces of tiles, pipes, wires or crushed rock that occur in or have been inserted in soil.
Forest	A plant community predominantly of trees and other woody vegetation usually with closed canopy.
Forestry	The act, occupation or art of farming and cultivating forests and systematic utilization, reproduction and improvement of the productive capacity of trees in masses, including the planting and culture of new forests.
Formula (mixed fertilizer)	The amount and grade of fertilizer materials used in making a mixed fertilizer.
Formulation	A mixture of an active pesticide chemical with carriers, diluents, or other materials-usually used to facilitate handling.
	A mixture of an active pesticide (herbicide) chemical with carriers, diluents or other materials; usually to facilitate handling.

Term	Terminology
Fossil	Any impression, natural or impregnated remains, or other trace of an animal or plant of past geological ages which has been preserved in the earth's crust.
Fouling crops	Whose cultural practices allow the infestation of weeds intensively?
Foundation seed	The second link in the certified seed chain produced from breeder seed and handled in such a way as to ensure genetic identity and varietal purity.
Free board	It is the vertical distance between the highest water level anticipated in the design of an irrigation structure (reservoir dam, canal, channel etc.) and the top of the retaining bank.
Free flow	It is a condition under which the rate of discharge is solely dependent on the length of crest and depth of water in the converging section of the partial flume. In case of weirs the criterion is the depth of water at 4 H (crest height) distance away from the weirs. There is no back pressure of water in a free-flow condition.
Free groundwater	Ground water in the interconnected interstices in the zone of saturation, it extends down to the impervious barrier, and moves under the influence of gravity in the direction of the slope of the water-table.
Freezing (frost) injury	Plant tissues are killed or severely injured when they are exposed to temperature low enough to cause ice formation in them.
Freezing point	Freezing point of a liquid is a characteristic temperature at which the liquid and the solid states of the substance can stay in equilibrium under a pressure of two atmospheres.
Frequency distribution	Refers to data classified on the basis of some variables, which can be measured on a scale.
Friability of soil	It characterizes the case of crumbling of soils. The moisture range in which soils are friable is also the range in which conditions are optimum for tillage.
Friable	A consistency term pertaining to the ease of crumbling of soils.

Term	Terminology
Frost damage	Damage caused by low temperature.
Frost day	Day with a minimum temperature less then 0°C.
Frost injury	Injury to the crop plants caused by temperature below the freezing point of water (0°C). Crops can be protected against frost injury by spraying water during the time when actually frost is occurring or by applying irrigation not long before a possible night frost.
Frost point	Maximum temperature of formation of frost by sublimation of atmospheric moisture on a cooled polished surface.
Frost-free season	A climatic record derived from the number of days during which the temperature is continuously above 0°C. For agricultural purposes, it is taken as a period between the last killing frost of spring and the first killing frost of autumn.
Fruit	The ripened ovary or group of ovaries together with other adhering structures.
Fruit drop	Any substantial dropping of immature fruit or of fruits as it approaches maturity.
Fuel	Any substance that can be burnt to produce heat. Sometimes includes materials that can be fissioned in chain reaction to produce heat.
Fuel efficiency	The ratio of the heat produced by a fuel for doing work to the available heat of the fuel. The efficiency is determined by the non-heat-forming materials in the fuel and the net work-producing heat which is developed by the fuel.
Full bloom	The plant at the period at which most of its flowers are blossoming.
Full ground cover	Soil covered by crop canopy approaching 100% when looking downward.
Fulvic acid	A term of varied usage but usually referring to the mixture of organic substances remaining in solution upon acidification of a dilute alkali extract of soil.

Term	Terminology
Fumigant	Chemical used in the form of a volatile liquid or a gas to kill insects, nematodes, fungi, bacteria, seeds, roots, rhizomes or entire plants. Usually applied in an enclosure of some kind or in the soil.
	A volatile liquid or gas to kill insects, nematodes, fungi, bacteria, seeds, roots, rhizomes or entire plants; usually applied to soil or in an enclosure of some kind.
Fumigation	The process of fumigating seed lots by celphos, EDB etc to kill storage insect pests.
Functional allelopathy	The substance may be released as a precursor and may be changed into an. active substance to inhibit growth.
Fungicide	A substance which kills fungi.
Fungus	The thallophyta lacking photosynthetic pigments. The plant body is made up of simple filaments.
Furrow	The trench formed by a tool in the soil during operation. Furrow crown: The peak of the turned furrow slice.
Furrow face	The vertical side of the furrow made in the soil, away from which the turned soil is thrown.
Furrow irrigation	A method of applying irrigation water to fields or orchards by small ditches or furrows which lead from the supply ditch.
Furrow planting	Planting in the bottom of furrows.
Furrow slice	The mass of soil cut, lifted and thrown to one side of an implement.
Furrow sole	The bottom of the furrow on which the plough bottom slides.
Furrow wall	It is an undisturbed soil surface by the side of a furrow.
Fused tricalcium phosphate	A product composed chiefly of the alpha form of the compound represented by the formula $Ca_3(P_2O_5)_2$. It is obtained when rock phosphate containing 5–10% silica is fused and the melt quenched.

Term	Terminology
Fuzz	The unmarketable and non-separable lint or fiber attached with cotton seeds.
Gamete	A sex cell which, after union with another develops into a new plant.
Gamma-rays	Electromagnetic radiation having its origin in an atomic nucleus.
Garden agriculture	Crops grown on land more or less adjacent to settlements; usually vegetables etc. rather than staples; dependent on organic wastes from settlement for nutrient inputs.
Gas constant	The constant factor in the equation of state for perfect gases. The universal gas constant is $R = 8.314 \times 10^7$ egr mol^{-1} ^{0}K^{-1}.
Gaseous fertilizer distributor	A machine for applying mineral fertilizer in gaseous form, which injects the fertilizer under pressure into the ground.
Gasification	It refers to conversion of biomass wood into medium Btu gas or conversion of coal into a high Btu synthetic natural gas under conditions of high temperatures and pressures. In gasification, biomass reacts with steam and oxygen resulting in gas consisting mainly of CO and H_2 which can be burnt as a fuel or used to produce fuels like methanol, H_2 and NH_3.
Geiger counter (nuclear)	An ionization chamber operating in the Geiger region, and is used in measuring activity present in a radioactive sample.
Gene	A hereditary germinal factor or unit in the chromosome which carries a hereditarily transmissible character, consisting of a group of based pairs within the DNA.
Generic	Pertaining to genus, a taxonomical category between family and species which includes closely related species.
Genesis soil structure	The causes and method of formation of the structural units or aggregates of soil.
Genetic erosion	Gradual disappearance of various forms of a cultivated species and of its wild relatives.

Term	Terminology
Genetics	The science of heredity including the study of its chemical foundation, its development, expression and its bearing on variation, selection, adaptation, evolution, breeding and the activities of man.
Genotype	The hereditary make up of an individual plant organism which, with the environment controls the individuals, characters.
	The organism or an individual that possesses a separate and distinct set of genetic constitution or a set of genes.
Genus	A group of closely related species. In a scientific name the genus is the first of the two names given for an organism.
Geological erosion	Erosion due to geologic process as distinguished from accelerated erosion.
Geometric mean	It is the root of the product of n items or values.
Geomorphology	The study and interpretation of land-forms. Geotropism: A growth movement in response to gravity. Roots are positively geotropic and shoots are negatively geotropic.
Germicide	An agent capable of killing germs usually pathogenic micro-organisms.
Germination	It is the resumption of growth of a seed usually reorganized by rupture of the seed-coat or spore wall and appearance of the radical and plumule from the seed.
	The period during which physiological processes are initiated in the seed leading to the elongation of cells and the formation of new cells, tissues and organs, morphologically observable as radical protrusion through the seed coat.
Gibberellins	Growth hormones, not identical with auxins, which markedly increase the elongation of stems in many plants, they also affect other processes.
Ginning (cotton)	The process of separating lint from seed cotton in ginning factory.

Term	Terminology
Ginning percentage	The weight of cotton lint obtained from seed cotton expressed in term of percentage of seed cotton or percentage of lint obtained from 100 unit of seed cotton by weight.
Glacial drift	A general term for the rock debris that has been transported by glaciers and deposited either directly from the ice or from the melt water with the melting of the glacier.
Gleization	The reduction of iron under anaerobic 'waterlogged' soil condition, with the production of bluish to greenish gray matrix colours, with or without yellowish-brown, brown and black mottles and ferric and magniferous concretions.
Gleying (rice culture)	Maintaining soil layer under reduced condition. This may occur due to watertable being nearer to the surface or maintaining water on the soil surface during most of the rice-growing season.
Global radiation	The total of direct solar radiation and diffused sky radiation received by a unit horizontal surface. Total earth surface receives $860+10^{18}$ kCal.
Glucose	A simple sugar ($C_6H_{12}O_6$) also known as dextrose.
Glucosides	Nitrogenous compounds that yield sugar and some other substances, usually aromatics on hydrolysis.
Glucosinolates	Sulphur containing substances that are broken down by the enzyme myrosinase to give bitter testing, toxic and goitrogenic compounds in the meal and 4 major alkenyl (aliphatic) glucosinolates are gluconapin, glucobrassicanapin, progoitrin and neoglucobrassicin and two indol (aromatic) glucosinolates are glucobrassicin and neo-glucobrassin. They are all hydrolised by endogenous myrosinase enzymes released from parenchymatous cells when these are crushed.
Glumes	A pair of bracts at the base of a spikelet in grasses.
Glutelins	Proteins those are insoluble in neutral solutions but soluble in weak acidic or basic solutions. These proteins are mostly formed in cereal grains, for example glutenin from wheat and oxyzenin from rice.

Term	Terminology
Gluten	An elastic substance that gives adhesiveness to dough. It is formed when the protein in flour, especially those in wheat flour absorb water. It assists in giving shape to the cooked product as it coagulates upon heating.
Glycolipids	Substituted lipids containing carbohydrates and nitrogen.
Glycolysis	A series of biochemical reaction in respiratory process in which hexose sugar is converted to pyruvic acid with the net gain of 2 ATP molecules. It is an anaerobic process.
Glycoproteins	In addition to simple proteins, glycoproteins contain some carbohydrates.
Gossypium arboreum	A diploid old world cotton species with n=13 chromosomes; which is widely grown in rainfed areas. It includes perennial or annual shrubs, 2-10 ft high. The centre of origin is believed to be Bengal region of Asia. Ginning percentage ranges from 24-28.
Gossypium barbadense	A tetraploid New World cotton species with n=26 chromosomes. The centre of origin in tropical South America. Ginning percentage ranges from 32 to 34.
Gossypium herbaceum	A diploid old world cotton species with n=13 chromosomes, constitutes a fairly large percentage of medium-staple cotton grown in India. Compared to arboreum cotton, these are of longer duration and better suited to deep and retentive soils but the upper limits of yield, lint length, fiber weight, fiber maturity in this species are inferior to arboreum types. Ginning % varies from 27 to 36.5.
Gossypium hirsutum	A tetraploid New World species with n=26 chromosomes. The centre of origin is known to be Central America. Ginning percentage varies from 33 to 37.
Gossypol	A phenolic pigment in cotton seed that is toxic to some animals.
Grade (fertilizer)	The minimum guarantee of plant nutrient content expressed as whole numbers in terms of total nitrogen (N), available phosphoric acid (P_2O_5) and water-soluble potash (K_2O).

Term	Terminology
Graded terrace	A terrace having a constant or variable grade along its length.
Grading	The process of sorting grains into different lots conforming to certain predetermined standards.
Grain	An indehiscent simple fruit of a grass in which the seed coat is united with the pericarp.
Grain cleaner	A machine to remove foreign matter from grain mass.
Grain cleaning	The cleaning that is based on function of air velocity, which is used to separate grains from trash by weight, density and resistance.
Grain development	A continuous process in which grain undergoes distinct-changes before it fully matures.
Grain drying	Drying the harvested grain usually in sun to reduce the grain moisture content to an acceptable level. Drying process is basically the transfer of heat by converting the grain water to a vapour and transferring it to the atmosphere.
Grain grader	A machine used for grading the grain.
Grain legumes	These crops belonging to the family Leguminosae, which are grown for their edible seeds.
Grain processing	The process of upgrading grains for improving their marketability, storability and suitability for human consumption.
Grain separator	A machine to remove impurities from grain or other seeds and to sort them into two or more fractions.
Grain shattering	A term used to describe the loss of grain or seed from the ear or spike while crop is still standing in the field.
Gram (Pulse)	Name of a pulse crop or grain legume also known as Bengal gram or chickpea (*Cicer arietinum* L.).
Gram atom	Synonym for gram atomic weight. The mass of an element in grams, numerically equal to the atomic weight.
Gram equivalent	The mass of a substance which will react with or is otherwise equivalent to one gram atom of hydrogen.

Term	Terminology
Gram molecular weight/ gram mole	Mass of substance in grams numerically equal to the molecular weight.
Gram positive	Bacteria that are stained following the gram staining process.
Gram stain	It is a differential staining by which bacteria are classed as gram-positive or gram-negative depending on whether they retain their primary colour (crystal violet) when subjected to treatment with a decolourising agent.
Gram-negative bacteria	Bacteria that are not stained following the gram-staining process.
Grand growth	The period of maximum enlargement of cell, tissue organ, or organisms. The enlargement starts slowly immediately after differential, increases to a maximum rapidly and finally falls to zero.
Granular	A dry formulation consisting of discrete particles generally less than 10 cubic millimeters and desisted to be applied without a liquid carrier.
Granular fertilizer	A fertilizer composed of particles of roughly the same composition and about one-tenth of an inch in diameter. This kind of fertilizer is superior in efficiency compared to the fine or powdery fertilizer, due to ease in handling and less fixation in soil.
Granular structure	Soil structure in which the individual grains are grouped into spherical aggregates with indistinct sides. High-porous granules are commonly called crumbs. A well granulated loamy soil has the best structure for most crop plants.
Granulation	Techniques using a process such as agglomeration, accretion or crushing, to make a granular fertilizer.
Granules (pesticide)	Granular formulation consists of free flowing grains, of inert materials mixed or impregnated with an insecticide. These are the easiest formation to apply.
Granum	One of the disclike bodies containing chlorophyll which is present in a chloroplast.

Term	Terminology
Grass	Botanically, any plant of the family Gramineae is called grass. The term grass in the context of feeding livestock and grassland agriculture is not limited to the narrow botanical sense alone, but also includes their common associates of the legume family. However, the term does not include cereals when grown for grain.
Grass tetany (hypomagnesemia)	Conduction of cattle and sheep marked by tetanic stagers, convulsions, coma and frequently death; characterized by a low level of blood magnesium.
Grassed water way	A natural or constructed water way, usually broad and shallow, covered with erosion-resistant grasses used to conduct surface water from crop land.
Grassland	The land on which graminaceous species represent the dominant, if not the exclusive vegetation. Grassland is intermediate in status between forest or woodland on one hand and the desert on the other.
Grassland ecosystem	It pertains to the ecosystem of grasslands and pastures.
Grassland farming	Farming system that emphasizes the importance of grasses and legumes in livestock and land management in which the legumes are the keystones and grasses· form the backbone of the grassland farming.
Gravitational head of water	Gravitational head of water in soil at a given point is the elevation of the point with respect to an arbitrary datum or reference point.
Gravitational water	It is the excess water above field capacity in a soil moving down under the influence of gravity.
Grazing	The eating or partial defoliation of any kind of standing vegetation by domestic livestock or other animals.
Grazing capacity	Number of animals a given pasture will support at a given time or for a given period of time.
Grazing intensity	The number of animals per unit area of grazing land.
Grazing pressure	Number of animals per unit area of available forage.

Term	Terminology
Grazing value	The worth of a plant or cover for livestock determined by palatability and nutritional value as excellent, fair, good or poor.
Grazing-selective	The preferential and sometimes excessive grazing of certain plants in mixed pasture.
Greenhouse	A house covered with glass used for the protection and cultivation of plants.
Green manure	Any crop or plant grown and ploughed under when succulent to improve the soil especially by addition of organic matter.
Green manure *in situ*	A practice of ploughing or turning into the soil un decomposed green-manure crops in the same field where the crop is grown.
Green manure trampler	An implement used to trample and press the green manure crop in the fields.
Greenhouse effect	The effect of short-wave radiation can easily pass through the atmosphere to reach the earth's surface while the outgoing radiation from the earth are not permitted to pass through the atmosphere, especially when it is cloudy or the ability of the atmosphere which admits most of the in isolation and prevent back radiation.
Green-leaf manuring	This refers to turning under of green leaves and tender green twigs collected from shrubs and trees grown on bunds, wastelands and nearby forest areas. The common shrubs and trees useful for this purpose are glyricidia *(Gliricidia tnaculata), Sesbania speciosl, karanj (Pongamia bin nata),* Subabul *(Leucopenia leucocephala).*
Green-manure crop	A crop grown and ploughed down while still green to add dry matter and nitrogen to the soil.
Gross cropped area	The total area covered with crops during an year. When three crops are grown on same land during one year, the same land area is counted thrice.
Gross irrigation requirement	The total amount of water applied through irrigation including losses during conveyance, distribution and application of water in the fields etc.

Term	Terminology
Gross minimum temperature	It is the minimum temperature indicated by a thermometer (namely alcohol-in-glass type) whose bulb remains in contact with top of the grass blades of short turf. This instrument gives an indication of ground frost.
Gross productivity	Total productivity of a system or plant per unit area and per unit time.
Gross return	Total income from the farm, by virtue of sales of entire farm produce.
Ground cover	Protective vegetation such as provided by a solid seeded stand of forage species or the residue from the previous crop.
Ground frost	Temperature of the upper layer of the ground or soil is less than 0°C.
Ground phosphate rock	Material obtained by grinding naturally occurring phosphate rock to fineness meeting relevant legislation or accepted custom.
Ground state (nuclear)	The lower energy level of a nucleus, atom or molecule.
Groundnut decorticator	A machine is used to separate groundnut kernels from pods.
Ground water	Subsurface water in the zone of saturation that is free to move under the influence of gravity.
Growing degree-day	Sum over the growing season of a crop of the difference between the daily temperatures and a reference temperature.
Growing period	The duration, in days, of the period when both temperature and soil moisture permit crop growth (crop growing season, growth cycle).
Growing season	Used in a general way, not as a technical term, to refer to the period of the year when crops are grown.
Growth	An irreversible process in which there is an increase in size or dry weight or in volume of an organism.

Term	Terminology
Growth analysis	Method of analysing growth through plant sampling technique. By this method observations are recorded periodically on various plant characteristics (height, flowering time, ear-bearing tillers etc.) and also sampled at varying intervals (weekly or fortnightly) when the different parts such as stem, root, leaf, fruit, seed etc. are separated, weighed and measured. From these the physiological parameters such as growth rate are calculated and these are further analysed in term of contributing component parameters like net assimilation rate and leaf area.
Growth correlation	The different parts of a plant do not grow in isolation but follow an orderly pattern due to control of one part of the plant over the other. This controlling effect exerted by one part over the other is known as correlative effect.
Growth curve	A graph in which the size of some plant characteristics (height, weight, leaf area etc.) is plotted against time or age. The rates of growth of any single cell or mass of cells are sigmoid (S-shaped), slow at first, increase with time to a maximum and then all off at about the same rate to zero. The period of maximum growth is called 'grand period' of growth.
Growth factor	A specific substance that must be present in the growth medium to permit cell to multiply.
Growth medium	Any material such as soil, peat, etc. used as a support for plant roots, that has a capacity for water retention and which may contain added and or naturally occurring nutrients.
Growth Regulator	An organic substance effective in minute amounts for controlling or modifying plant processes.
Growth stages of cerealcrops	(1) Tillering stage-additional shoots are developing from the crown, (2) jointing stagewhen stem internodes are elongating, (3) boot stagewhen leaf-sheath swells due to the growth of developing spike or panicle, (4) heading stagewhen seed head is emerging from the sheath.
Growth substance	A growth regulating chemical produced by a plant or a synthetic one with similar activity even at very low concentrations.

Term	Terminology
Growth-regulator (plant)	Organic substances which in minute amounts may participate in the control of growth processes. In plant 4 types have been recognized to date: auxins, gibberellins, kinins and inhibitors.
Growth-retardant	A chemical (such as CCC) that selectively interferes with normal hormonal promotion of growth but without appreciable toxic effects.
Guano	Includes many materials which vary in source, composition and readiness for use. It may be (1) bat guano found in caves; (2) peruvian guano the accumulated excrement of sea birds found in peru; (3) fish guano, whale guano, sheep guano, goat guano; and (4) phosphatic guano of various kinds. The nitrogen content varies from 0.4 to 9.0% and total P205 from 12 to 26%.
Guard cell	One of two epidermal cells which enclose a stoma.
Guard cropping	Means growing main crop in the centre surrounded by hardy and thorny crop, such as safflower around pea, mesta around sugarcane and sorghum around maize.
Gully erosion	Removal of soil by running water, with formation of channels that cannot be smoothed out completely by normal cultivation.
Guttation	Exudation of water usually through structures called hydathodes present at the tips of veins of leaves. This occurs in plants growing under moist and humid conditions.
Gymnosperm	A seed plant in which the seeds are not enclosed in ovaries. Pine, fir, spruce and other conifers are the examples.
Gypsophyte	A plant that grows on gypsum soils.
Gypsum	A hydrated calcium sulphate ($CaSO_4.2HP$). The commercial material contains varying amounts of impurities and is used as a soil amendment.
Gypsum block	Made up from plaster of Paris with the help of two electrodes parallel to each other in a medium of gypsum block. It is a resistance unit which is used for measurement of soil moisture in 'situ'. The resistance reading is about 400-600 ohms at field capacity and 50,000-75,000 at wilting point.

Term	Terminology
Gypsum requirement	The quantity of gypsum or its equivalent required to reduce the exchangeable sodium percentage of a given alkali soil to an acceptable level.
Gyro sifter	A machine in which grains of smaller size are separated by passing them over gyrating screens. Small-size grains tend to get collected at the centre while the larger ones move towards the periphery and get separated.
Habitat	A specific kind of living space or environment where an organism lives or could be found.
Hail	Solid precipitation in the form of ice pellets from cumulonimbus clouds due to severe up draughts. Hail diameter ranges 5-50 mm and has onion-like cross section.
Hailstorm	A storm often prolonged and severe, consisting largely of hail or frozen raindrops ranging in diameter from 5 to 10 mm or more, the ice particles are precipitated either separately or as aggregates of irregular size and shape.
Hair hygrometer	Hygrometer in which sensing element is a strand or strands of hair, the change of its length is a function of humidity of the air.
Half-life (nuclear)	The time it takes for any initial amount of a radio nuclide to be halved by radioactive decay.
Half-thickness or half-value layer (nuclear)	The thickness of any absorbing material required to reduce the intensity of a beam of radiation by half.
Halophytes	Plants tolerant to high soil salinity conditions.
Haploid	Having one set of chromosomes, as in gametes.
Hard pan	It is a hard and impermeable layer formed in the soil profile by accumulation of materials such as salts, clay etc., and which impedes drainage.
Hard seed	Seeds that have a seed-coat imperviolls to water or oxygen required for germination.
Hard water	Water which contains certain minerals, usually calcium and magnesium sulfates, chlorides or carbonates in solution, to the extent of causing a curd or precipitation rather than lather, when soap is added.

Term	Terminology
Hardening	Increase in the tolerance of a plant tissue to stress or adverse conditions.
Harmonic mean	It is the inverse of the arithmetic mean of the, inverse of given number.
Harrow	A secondary tillage implement that cuts the soil to a shallow depth for smoothening and pulverizing the soil as well as to cut the weeds and to mix materials with the soil.
Harrowing	A secondary tillage operation which pulverizes smoothens and packs the soil in seed-bed preparation and or controls weeds.
Harvest diversity index	It measures the multiplicity of crops or farm products which are planted in a single year by computing the reciprocal of sum of squares of the share of gross revenue received from each harvest in a single year.
Harvest index (HI)	Yield of the plant parts of economic interest (economic yield) as percentage of total biological yield in terms of dry matter.
Harvest maturity	It refers to a developmental stage which generally occurs seven days after physiological maturity during which loss of moisture from the plant occurs.
Harvesting	The operation of cutting, picking, plucking or digging, or combination of these for removing the useful part or economic part from the plants.
Haulm	Stem or stalk of a plant such as portative peas etc. after harvesting.
Haustorium	The major organ of parasitic weeds for attachment and penetration of the host tissue.
Hay	Hay is feed produced by dehydrating green forage to a moisture content of 15% or less, so that the biological processes do not proceed rapidly enough to build up heat.
Haylage	Product resulting from ensiling forage with 35-50% moisture under anaerobic conditions. Compared to hay, it greatly reduces the time, involved in field curing and the associated field risk.

Term	Terminology
HCN poisoning	Forages such as sorghum, Sudan grass, lotus species and white clover contain naturally occurring cyanogenic glycosides. Large amounts of HCN (hydrocyanic acid) may be released through cyanogenesis process in a relatively short period of time when plant tissue is injured by cutting or grazing. Ruminants are considered susceptible to HCN poisoning because rumen micro flora are also capable of enzymatic breakdown of the glycosides. HCN is readily absorbed into blood and is distributed to all tissues of the body. HCN acts by forming an inactive complex with cytochrome oxidase, a key enzyme in cellular respiration. HCN poisoning, therefore, constitutes asphyxiation at the cellular level. Levels of HCN greater than 20 mg/100 g dry plant tissue are considered dangerous. Generally, the HCN potential is greatest in new plant growth particularly after environmental stresses such as drought and frost, or after application of high levels of nitrogen fertilizer.
Heading	Emergence of the panicle out of the flag-leaf sheath is called heading. It occurs after booting.
Headland	A barrier of bushes, shrubs or small trees growing close together in a line.
Heat capacity	The ratio of heat absorbed by a system to the corresponding temperature rise.
Heavy soils	A term for clayey or fine-textured soils.
Heavy water (D_2O)	Water in which hydrogen has been replaced by deuterium (hydrogen isotope of mass 2).
Hectare	It is a unit of field area consisting 10,000 m^2 or having an area of 2.471 acres of land.
Hectare-meter water	Water column having vertical depth of one m stored over one ha area subject to no leaching, percolation, seepage or evaporation of water. It is equivalent to 10,000 m^3 water or 10,000 tonnes of water.
Hekistotherm	A plant which grows under very low temperatures, generally in alpine areas.
Heliograph	It is a sunshine recorder which records the actual hours of bright sunshine.

Term	Terminology
Heliotrophic	Turning towards the sun.
Helophytes	Plants of sun-loving species require intense light for normal development.
Hemicellulose	Group of substances consisting of araban, xylan and polyuronides which are much less resistant to chemical agents than cellulose.
Herbaceous plant	A vascular plant that does not develop woody tissue.
	A vascular plant without woody tissues.
Herbage	Leaves stem and other succulent parts of forage plants upon which animals feed.
Herbarium	A collection of plant specimens that have been taxonomically classified, pressed, dried and mounted on sheet of herbarium paper.
Herbicide	A chemical used for killing or inhibiting the growth of unwanted plants.
	A chemical used for killing or inhibiting the growth of plants; phytotoxic chemical (from Latin *herba,* plant *andcaedere,* kill).
Herbicide mulch	Loosely woven cloth or other like material, impregnated with herbicide to fit the crop inter-rows for controlling young, germinating weeds there.
Herbicide residues	It is a part of the herbicide left behind unused and undetoxified after their use in the soil.
Herbicide resistance	The traits or quality of a population of plant within a species or plant cells in tissue culture of having a tolerance for a particular herbicide that is substantially greater than the average for the species and that has developed because of selection naturally occurring tolerance by exposure to the herbicide through several reproductive cycles.
Herbicide rotation	System of using different kinds of herbicides in recurrent succession on same land.
Herbicide selectivity	Refers to the phenomenon where a chemical kills the target plant species in a mixed plant population without harming or only slightly affecting the companion plants.

Term	Terminology
Herbicide therapeutics	Herbicide molecules possessing certain plant disease suppression effects.
Herbigation	It refers to application of herbicide with irrigation water.
Heredity	Measure of the degree to which a character is controlled by the genotype compared to the environment.
Hermophrodite	A plant having both male and female parts in the same flower
Heterogygous	Not true-breeding for a specific hereditary character.
Heterotroph	An organism that cannot make its own food and hence obtains both organic and inorganic materials from the environment.
Heterotrophic	Capable of deriving energy for life processes only from the assimilation of organic carbon compounds and incapable of using carbon dioxide as the sole carbon source for cell synthesis. Contrast with autotrophic.
Hidden hunger	It is a term used to describe a plant that shows no obvious symptoms yet the nutrient content is not sufficient to give profitable yield.
High analysis fertilizer	A fertilizer containing not less than 25% of the major plant nutrients, namely nitrogen, phosphorus (as P_2O_5) and potassium (as K_2O).
High inputs	Methods based on advanced technology and high capital resources; fertilizers at levels of maximum economic return; chemical weed and pest control at advanced technical levels; modern method of mechanization are employed to maximize yields or economic return.
High level equilibrium farming	Farming involves sole crops and high cost inputs as followed in the developed countries.
Hill agriculture	Farming system practiced on high-altitude terraced lands or valleys with rolling landscape.
Hill dropping	A method of sowing in which seeds are dropped at fixed spacing and not in a continuous stream. Thus the spacing between plant to plant in a row is constant.

Term	Terminology
Hill placement	This refers to applying fertilizer either in bands or near the plants. This method is usually employed when relatively small quantities are to be applied. Nitrogenous fertilizer is applied in pinches at the base of each plant of cotton, maize, cauliflower, cabbage, brinjal, tomato and other vegetables by this method.
Hilum	The scar on a seed-coat where the funiculus (stalk) is attached. The bright centre of a starch grain.
Histogram	A bar diagram representing a frequency distribution.
Histology	Study of tissues, either of plant or animal.
Histosols	Soils formed from materials high in organic matter, histosols with essentially no clay must have at least 20% organic matter (12% C) by weight (about 78% by volume). This proportion increases to 30% (85% by volume) in soils with at least 50% clay.
Homeostasis	The tendency of the biological system to resist change and to remain in a state of equilibrium.
Homogeneity test	A test to measure variation within a seed lot. Homozygous: Having identical genes at the same locus in each member of a pair of homologous chromosomes.
Hoof and horn meal	A product resulting from processing, drying and grinding of hooves and horns and it contains 13-15% nitrogen.
Horizon (soil)	A layer of soil approximately parallel to the land surface with more or less well-defined characteristics.
Horizontal revolution in agriculture	Increased land use by expanding cultivated land area through the utilization of fallow and marginal lands and reclaiming culturable wastelands, thereby increasing land-use intensity.
Hormone	An organic substance produced naturally in higher plants, controlling growth or other physiological functions at a site remote from its place of production and active in minute amounts.
	Growth-regulating substance occurring naturally in plants or animals; more correctly called growth regulator, not hormone.

Term	Terminology
Horse power	A unit of power, equivalent to 33,000 foot pounds per minute, or 550 foot pounds per second (mechanical horse power) or 746 watts.
Host	An organism or a plant that harbours a parasite such as disease producing fungus or an insect that feeds upon it.
Hub crop	The crop in sequential cropping system which has the greatest comparative advantages over other crops, *e.g.,* vegetables/truck crops cultivated near city markets.
Hull (rice)	The rice caryopsis is surrounded by a hull (husk) composed of two modified leaves, the palea and larger lemma.
Huller	A machine which is used for hulling. While removing the hull, the machine removes part of the bran also.
Hulling	It is the process of removing hull or husk and bran in one operation.
Humic acid	A term of varied usage but usually referring to a mixture of indefinite composition of dark-coloured organic substance precipitated upon acidification of a dilute alkali extract of soil.
Humid tropics	The tropical areas with excessive moisture and are characterized by isothermal conditions which facilitate year round crop production.
Humidity	A general term indicating the amount of water vapour in the air.
	Moisture or dampness in the air.
Humidity (relative)	The ratio of the weight of water vapour in a given quantity of air to the total weight of water vapour that air is capable of holding at the temperature in question; expressed as per cent.
Humidity index	Water surplus as a percentage of annual potential evapo-transpiration in dry climates.
Humification	Processes of organic decomposition leading to the formation of humus.

Term	Terminology
Humin	It is that part of the organic matter not dissolved upon extraction of soil with dilute alkali.
Humus	The well decomposed, more or less stable part of the organic matter of the soil.
Hybrid	The first generation progeny or offspring from a cross of different varieties or strains or inbred lines.
Hybrid vigour	The enhanced vigour of progeny resulting from the crossing of two different inbred plants.
Hybridization	A method of crop improvement in which two or more plants of unlike genetical constitutions differing in one or more characters are crossed together.
Hydathode	A pore generally at the tip of a vein, capable of exuding liquid water.
Hydration	The association of one or more molecules of water with an ion, molecule or micelle.
Hydraulic conductivity	A proportionality factor 'K' in the Darcy's equation which states that the effective flow velocity 'V' in a porous medium is proportional to hydraulic gradient, $(h_1 - h_2)/L$.
Hydraulic equilibrium	Hydraulic equilibrium is the condition for zero flow rate of water film in soil.
Hydraulic gradient	The decrease in hydraulic head per unit distance in the soil in the direction of the greatest rate of decrease of hydraulic head.
Hydraulic head	An elevation with respect to a standard datum at which water stands in a riser or a manometer connected to a point in question in the soil.
Hydraulic radius	It is the ratio of the volume to the surface of the pore space or the average ratio of the cross sectional area of the pores to their circumferences.
Hydrograph	An instrument for recording automatically and continuously the variations of the relative humidity in the atmosphere.
Hydrography	The science that deals with the waters of the earth's surface, particularly with reference to their physical features, position, volume etc. and the precipitation charts of seas, lakes, rivers, contours of the sea bed, shallows, deep currents etc.

Term	Terminology
Hydrologic equation	The water inventory equation (inflow = outflow + storage) which expresses the basic principle that during a given time interval the total inflow to an area must equal the total outflow plus the net change in storage.
Hydrological cycle	The continuous circulation of water among the hydrosphere, atmosphere and lithosphere.
Hydrology	The science that deals with water especially in relation to its occurrence in wells, lakes, streams etc. and as snow including its uses, conservations, control and discovery.
Hydrolysis	Breakdown of a substance with the addition of water.
Hydrometer	An instrument used to find the specific gravity of a liquid.
Hydrophilic	The character of substances having greater affinity for water and other polar solvents.
Hydrophytes	Plants growing habitually in water or in very wet soils where at least periodically oxygen becomes deficient as a result of excessive water content.
Hydroponics	The growing of plants in water solutions of essential nutrients.
Hydroscopic coefficient	It is the amount of moisture in dry soil when the same is in equilibrium with some standard relative humidity near a saturated atmosphere (98%), expressed in term of percentage on the oven dry soil.
Hydrostatic pressure	The pressure in a fluid in equilibrium which is due solely to the weight of fluid above.
Hydrotropism	The growth movement in response to unequal distribution of water, as when roots bend towards moist soil.
Hygroscopic	A term used to designate materials that absorb moisture from the air, *e.g.,* urea.
Hygroscopic coefficient	It is the amount of moisture in dry soil when it is equilibrated with some standard relative humidity near a saturated atmosphere (about 98%) expressed in terms of percentage on oven dry weight basis.

Term	Terminology
Hygroscopic water	The amount of water which is absorbed by the soil from an atmosphere of saturated water vapour due to attractive forces on the surface of the particles.
Hygroscopicity of fertilizer materials	Some fertilizer materials absorb moisture from the air, which causes them to become sticky. Such hygroscopic fertilizers are calcium ammonium nitrate, ammonium chloride and urea.
Hygrothermograph	An instrument that automatically and continuously records both temperature and relative humidity on a single chart.
Hypocotyl	The part of the embryo or seedling below the cotyledonary node and above the root or the transition region connecting the stem and root.
	Portion of the stem of a plant embryo or seedling below the cotyledons.
Hypogeal germination	A type of germination in which the cotyledons remain below the soil surface.
Hypothesis	A speculative statement based on observation suggesting a possible cause and effect-relationship and forming the beginning point of an experimental study.
Hysteresis	It is the lag in one of the two associated processes or phenomena during reversion, in soil moisture absorption and disorbtion process.
Ideotype	Refers to plant type in which morphological and physiological characteristics are ideally suited to achieve high production potential and yield reliability.
Igneous rocks	Formed from molten material or magma by cooling and solidification.
Illite (2 : 1 non-expanding type)	Its structure is similar to montmorillonite group but has much longer particles with considerable interlayer potassium. It belongs to hydrous mica group.
Illuviation	The process of deposition of soil material removed from one horizon to another horizon of the soil, usually from an upper horizon to a lower horizon in the profile.

Term	Terminology
Imbibition	The uptake of a liquid with swelling by a substance, such as seed, cellulose, agar or gelatin.
Imbibition pressure	An index of potential maximum pressure which can develop in an imbibant as a result of imhibitions.
Immature soil	A soil with indistinct or only slightly developed horizons because of the relatively short time it has been subjected to the various soil-forming processes. A soil that has not reached equilibrium with its environment.
Immobilization	The conversion of an element from inorganic to organic combination in microbial or plant tissues. This has the effect of rendering unavailable (and usually not readily soluble) an element that previously was directly available to plants.
Immunity	Natural or acquired resistance of a plant to active infection by the pathogenic micro-organisms or to the adverse effects of a particular parasite.
Impermeable (seed)	Impermeable seed is one whose seed-coat allows no passage through to water or gases.
Impervious soil	A soil through which air, water or plant roots penetrate very slowly if at all.
Implements	The equipment used on farm to carry out different farm operations either with the help of tractor or bullock power.
Importance value (IV)	The importance value of a species indicates the degree of dominance of species over the other species in a given sample plot.
Impoverished soil	Soil which has been continuously cropped and has had little or no care, so that it is incapable of supporting a satisfactory or economic crop.
Improved production technology	It is the use of improved varieties and associated improved crop-management practices followed on a farm.
Improved seed	The genetically and physically pure seed of an improved crop variety. Its different categories are nucleus seed, breeder's stock seed, foundation seed, registered seed and certified seed.

Term	Terminology
In situ	In the natural or original position or location.
In vitro	In glass, in test tubes, outside the organism, as digestion *in vitro.*
In vivo	In a living organism such as in the animal system.
Inbred	The progeny of either a single cross-pollination plant obtained by selfing or two closely related plants obtained by inbreeding.
Incipient Wilting	Small loss in turgidity during warm weather, even when soil is moist. It may not lead to any visible stress symptoms like rolling or folding of leaves and is usually reversible at night.
Income equivalent ratio (IER)	The ratio of the area needed under sole cropping to produce the same gross income as one hectare of intercropping at the same management level. IER is the conversion of LER into economic terms.
Income equivalent ratio (IER)	The ratio of the area needed under sale cropping to produce the same gross income as one ha of intercropping at the same management level. IER the conversion of LER into economic terms.
Incompatibility	Inability of the gametes of two organisms to unite and reproduce sexually or inability of one substance to mix with other without undesirable reaction, *e.g.,* with references to insecticides, herbicides, fungicides where incompatibility involves loss of efficiency on mixing.
Incorporating or mixing	Tillage operations which mix or disperse foreign materials, such as chemicals or plant residues into the soil.
Incremental cost: benefit ratio (ICBR)	A ratio of benefit to cost invested for obtaining economic returns in rupees per rupee' further invested for the costlier, valuable inputs imposed such as fertilizers, weedicides, pesticides, fungicides etc. for achieving the optimum, maximum yields of crops. ICBR is always compared with the next lower level of application of the respective input.
Indehiscent	Refers to a fruit which does not split open at maturity.

Term	Terminology
Indeterminate growth	The flowers are borne on lateral branches; the central stem shows continuous vegetative growth and blooming for a long period, as in cotton.
Indeterminate plants	Plants which flower when the days are either long or short.
Indica rice	All the rice varieties not included in the javanica or japonica subspecies are grouped together as indica and are cultivated all over the tropical Asia. In this group, the range of variation is very high. They are generally tall, vigorously growing, profuse tillering, late maturing, photoperiod sensitive, lodging susceptible and adapted to low-fertility conditions.
Indicator	A substance used to test for acidity or alkalinity of a solution by a colour change, *e.g.,* litmus and phenolphthalein.
Indicator plant	A plant which reflects either by its presence or character of growth, specific growing conditiob and like deficiency of plant nutrient, soil-moisture stress etc.
Indigenous	An organism that is native to a particular habitat, as distinct from one introduced from outside the area.
Indole acetic acid (IAA)	A naturally occurring growth regulator, the major auxin found in plants.
Induced mutations	The mutations artificially produced with the help of mutagens.
Induced radioactivity (nuclear)	Artificial radioactivity or that produced by nuclear reactions.
Induced state	The state of the plant in which it is capable of flowering. In case of photosensitive plants for flowering it is the stage after the exposure to requisite photoperiodic cycles leading to flowering.
Inductive statistics	Procedures employed to arrive at broader generalization or inferences from sample data to populations.
Inedible	A substance that is not fit for human consumption such as poisonous roots and plants.
Ineffective rhizobium	A rhizobium which although induces nodulation on the host plant but fixes little or no nitrogen.

Term	Terminology
Inert ingredient	Any material in a pesticide formulation which has no pesticidal action.
	Any ingredient in a formulation which has no pesticide (herbicidal) action.
Inert matter	In seed testing, it refers to all foreign matter, such as pieces of broken or damaged seeds, empty glumes, lemmas, paleas, soil, sand, chaff, stem etc., and all other matter not seeds.
Infiltration	It is a process of water entry into the soil generally (but not necessarily) through the soil surface and vertically downward. Water may enter the soil through the entire surface uniformly as under ponding or rain or it may enter the soil through furrows or crevices, or it may move up into the soil from a surface below under high water-table.
Infiltration capacity or soil infiltrability	It is the flux which the soil profile can absorb through its surface where it is maintained in contact with water at atmospheric pressure and there is no divergent flow at the borders.
Infiltration rate	It is the rate of water entry into the soil when the flow is non-divergent.
Infiltrometer	A device by which the rate and amount of water percolating the soil is determined by measuring the difference between the amounts of water applied and that which runs off.
Inflorescence	Arrangement of a plant's flowers on the stem; collective flowers of a plant, *e.g.,* umbel, receme, spike, panicle etc.
Infrared	The invisible part of the light spectrum immediately beyond the red, it is identical with radiant heat.
Infrared analyser	An apparatus used to measure either CO_2 or water-vapour content of an air sample. They depend on the differential absorption of radiation in two sample tubes with identical geometry, so that it clearly records vertical profile of the above parameters in: and above the crops or vegetation.

Term	Terminology
Infrared thermometer	Used to measure leaf and canopy temperature of plants. It collects the thermal energy emitted by the plant surface by fixed focus optics and converts into electrical signal by a sensitive detector which is electrically conditioned for reading in temperature units. It is quite fast in collecting energy at the speed of light.
Inherent resistance	Quality of a plant to resist attack of pest or disease probably associated with some particular characters of structure or composition.
Inheritance	The reception or acquisition of characters or qualities by transmission from parent to off spring.
Inhibition	Prevention of growth or multiplication of microorganisms.
Inhibitor	A chemical substance that acts to prevent a process from occurring.
Inhibitory polyculture	The interaction between crop species in an inter-cropping system has a net negative effect on all species.
Initial intake rate	Rate at which water will enter the soil when water is first applied; mm/hr.
Inoculate	To bring the pathogen into contact with the host, host organ or medium.
Inoculation	The artificial introduction of micro-organisms on or into a medium, living system.
Inoculum	The material containing viable micro-organisms and used for inoculation.
Inorganic nitrogen (soil)	Nitrogen in inorganic forms, *viz.*, exchangeable ammonium, nitrite, nitrate, fixed ammonium, nitrous oxide and elemental nitrogen.
Insolation	Amount of direct solar radiation incident per unit on a horizontal area at a given level.
Intake rate	The rate of water entry into the soil expressed as a depth of water per unit of time. This term involves no restrictions on area of application and divergence of flow in the soil.

Term	Terminology
Integrated farming system	Integration of various agricultural enterprises, *viz.*, cropping, animal husbandry, fishery, forestry etc. A judicious mix of anyone or more with cropping complements the cropping enterprise.
Integrated fertility management	Technologies for management of fertilizing the crops with the objective to produce optimum crop yields with minimum fertilizer inputs taking into consideration ecological and socio-economic resources available for supply to the crop under a given agro-ecosystem, without deteriorating the soil fertility, such as application of organic manures, green manures, blue-green algae and bio-fertilizers along with use of inorganic fertilizers for higher, yields and improving the soil productivity.
Integrated pest control	In this method a variety of technologies in a single pest management is used with the objective to produce optimum crop yield at a minimum cost taking into consideration ecological and socio-economic constraints under a given agro-ecosystem.
Integrated pest management	It is a broad ecological approach that minimises pest population below economic threshold level by employing all available pest control technique like mechanical, biological, chemical and crop management practices in compatible manners.
Integrated weed management	Application of many kinds of weed management technology in a mutually supportive manner.
	A weed management system that uses all suitable control methods in a compatible manner, to reduce weed population and maintain them at levels below those causing economic injury.
Intensive cropping	Maximum use of the land by means of frequent succession of harvested crops.
Inter flow	It is the lateral seepage of water in a relatively pervious soil above a less pervious layer. Such water usually reappears on the surface of the soil at a lower elevation.
Interaction effect	Measured by the failure of simple effect of one factor to remain consistant over the various levels of the other factor.

Term	Terminology
Intercropping	It refers to growing of two or more generally dissimilar crops simultaneously on the same piece of land, base crop necessarily in distinct row arrangement. The recommended optimum plant population of the base crop is suitably combined with an appropriate additional plant density of the associated crop and there is crop intensification in both time and space dimensions.
	Growing two or more crops simultaneously on the same field. Crop intensification is in both the time and space dimensions. There is intercrop competition during all or part of crop growth. Farmers manage more than one crop at a time in the same field.
Intercrops	The crop raised in an orchard or other widely spaced crops for increasing the income from the same piece of land, *e.g.,* short-duration vegetables, pulses, oilseeds etc.
Intercultivation (interculturing)	Soil cultivation performed in standing crop.
Interculture	Arable crops grown below perennial crops.
Interfaces	The contact surfaces between a solid and liquid or between two immiscible liquids.
Interference	Total adverse effect that plants exert on each other when growing in a common ecosystem; includes competition, allelopathy, biotic interference and other modifications detrimental to plant growth.
Intermediate belt	Zone that lies between the belt of soil water and capillary fringe.
Intermediate inputs	Methods practiced by farmers who follow the advice of agricultural extension services but who have limited technical knowledge and/or capital resources; improved agricultural techniques; inputs adequate to increase yields but not to achieve maximum yields or maximum economic return; some fertilizers (*e.g.,* 50-100 kg per hectare, combined weight of nutrients expressed as elements); possibly some use of chemical weed or pest control.

Term	Terminology
Intermediate plants	Those plants in which flowering occurs in a small range of intermediate photoperiods but will not flower under light period either shorter than the lower critical value longer than the higher critical value.
Internode	Internode refers to the space between two successive nodes on the main stem or branches of the plant.
Interplanting	All types of seeding or planting a crop into a growing stand. It is used especially for annual crops grown under stands of perennial crops. It defines as (i) in woodland, the setting out of young trees among existing trees or bushy growth; (ii) in orchards, the planting of the farm crops along the trees, especially when trees are too, small to occupy the land completely; and in cropland, the planting of several crops together on the same land like the planting of soybean with maize.
Inter-row cultivation	Weeding between the crop rows is done by bullock or with tools.
Inter-tilled crop	A crop planted in rows followed by cultivation between the rows.
Intra-specific hybridization	Hybridization between, two varieties of the same species.
Intrazonal soils	A broad soil order in genetic classification of soils. It embodies soils that reflect the dominance of relief or permanent material in their formation. Effects of microclimate on soil as induced by relief may be expressed in their profiles, *e.g.,* bog soils.
Intrinsic permeability	Intrinsic permeability is the factor k in the equation $V = k\ dgi/n$, where V = flow velocity, d = density of ' liquid, g = scaler value of acceleration due to gravity, t = hydraulic gradient and n = viscosity of fluid.
Inventory	It is a list of assets and liabilities which are claims or debts against the business; in other words it is a detailed list of farm properties, with value assigned.
Inverse nitrogen yield concept	According to Wilcox, the yield of the crop is inversely proportional to its nitrogen content.

Term	Terminology
Inversion	The condition in which there is abrupt rise instead of fall in temperature with height is called inversion.
Invert emulsion	One in which the water is dispersed in oil rather than oil in water. Oil forms the continuous phase with the water dispersed therein.
	The suspension of minute water droplets in a continuous oil phase.
	One in which water is dispersed on oil (instead of oil in water); oil forms the continuous phase with the water dispersed therein.
Invert sugar	A mixture of the simple sugars, dextrose and levulose produced by the inversions of sucrose (cane sugar).
Iodine number	A measure of the degree of unsaturation of a fat, being the amount of iodine that chemically reacts with fat. It is therefore proportional to the number of double (unsaturated) bonds linking the carbon atoms of the fat.
Iodine value	It is the measure of unsaturation of fatty acids and of their esters. The results are, commonly expressed as per cent iodine absorbed.
Ion	A charged atom or particle formed usually by the dissociation of a molecule.
Ion activity	The effective concentration of an ion in a solution or soil water system as determined from the electromotive force of a cell composed of a reference electrode, whose potential varies with the activity of the ion under investigation.
Ion exchange	A chemical process involving the reversible exchange of ions between ions in solution and ions bound to an insoluble substance or resin.
Ion pair	A positive ion and an electron negative ion produced by the action of ionizing radiation.
Ionization	The process of breaking the neutral molecule of a substance into positively and negatively charged particles.

Term	Terminology
Ionization chamber (nuclear)	An instrument for measuring radiation. It operates by measuring the ions produced by the radiation in a given volume and consists of two electrodes between which an electrical field is maintained to collect the charge.
Irrigable area	The proportion of the arable area of an irrigation project which is subject to irrigated farm use. It excludes lands required for non-productive uses.
	Land under existing or potential irrigation development which by reason of topography, quality of land and other characteristics is physically suitable for sustained irrigation and for which an adequate and suitable water supply can be provided at reasonable cost.
Irrigated dryland	Areas with well-drained soils where water does not accumulate and irrigation is used mainly for dryland crops such as wheat, maize, potato, and other cash crops. Dryland rice may also be grown with intermittent type irrigation.
Irrigated wetland	Areas where invariably a rice crop is grown with irrigation. Availability of irrigation water may be year round or seasonal. Other non-rice crops, particularly high-value crops, may be grown with irrigation in rice-based patterns.
Irrigation	The application of water to soil to assist in the production of crops especially during stress periods.
Irrigation efficiency	It is the ratio expressed in percentage of water stored in the root-zone depth of the soil to the water delivered in the field from the farm-supply source.
Irrigation interval	Time between the start of successive field irrigation application on the same field, in days.
Irrigation period	The number of days that can be allowed for applying one irrigation to a given design area during the peak consumptive use period of the crop being irrigated. It is the basis for designing the capacity of an irrigation project.

Term	Terminology
Irrigation requirement	It is a part of total water requirement of a crop, exclusive of effective rainfall and soil moisture stored in the root zone or that contributed from the shallow ground water-table.
Irrigation response	The rate of increase in crop yield per unit of increase in water applied.
Irrigation structure	Any structure or device necessary for the proper conveyance, control, measurement or application of irrigation water.
Irrigation water	Water which is artificially applied in the process of irrigation. It does not include precipitation.
Irrigation water requirement	The net irrigation water requirement divided by the irrigation efficiency.
Isobar (nuclear)	One of a group of nuclides having the same total-number of particles (neutrons = protons) in the nucleus but with these particles so proportioned as to result in different values of Z, *e.g.,* ^{3}H and ^{3}He.
Isobars	Lines of equal pressure.
Isobath (water-table)	A line on a map or ground connecting points at the same height or elevation above an aquifer or water table.
Isocline	A line or curve connecting the least cost combinations of inputs for all output levels is known as isocline.
Isohyet	A line drawn on a map connecting points with equal rainfall.
Isomer (nuclear)	One of two or more nuclides having the same number of neutrons and protons in the nucleus (same Z and A) but existing in different energy states.
Isophene	Line of constant date of appearance of certain stages of plant and animal life.
Isopleth	A line of equal value of a given quantity.
Isoputyledene diurea (IUBDU)	A reaction product of urea and isobutylaldehyde containing about 30% N in the fertilizer grade. A slow release N-fertilizer.
Iso quants	Refers to all the possible combinations of two inputs, physically capable of producing the same amount of output.

Term	Terminology
Isotach	A line of equal wind speed. Isotherm: Lines of equal temperature.
Isotone	Anyone of several nuclides having the same number of neutrons in the nucleus but differing in the number of protons.
	One of group of nuclides of the same element (same Z) having the same number of protons in the nucleus but differing in the number of neutrons, resulting in different values of A.
IW: CPE ratio	Based on the climatological approach, scheduling of irrigation water is done and the depth of irrigation water is fixed for different crops.
Japonica rice	These are rice varieties developed in Japan and are adapted for cultivation in the subtropical and warm temperate regions. The japonica varieties mostly have oval and round grains, the spikelets show distinctive pubescence near the apex and the panicles are compact and dense. Generally they are early maturing having relatively short, sturdy straw, resistant to lodging and respond well to added fertilizer. They are photoinsensitive and lack seed dormancy. The rice is somewhat sticky when cooked.
Javanica or bulu rice	A subspecies of sativa rice which includes a small number of varieties adapted to lower altitudes and are mainly found in Indonesia. Javanica rice is characterized by a straw, long panicle with awed grains, sparse tillering habit and long duration.
Jhum cultivation	The slash-and-burn type of shifting cultivation in the hill tracts of Bangladesh and Assam.
Juvenile phase	The phase of vegetative growth of a plant before which it cannot be induced to produce flowers. This period may vary from plant to plant, *e.g.,* rice plants have a juvenile phase of one month.
Kainite	A mineral composed of potassium chloride and magnesium sulphate ($KCl, MgSO_4 \cdot 3H_2O$) the crude potash ore sold as kainite has varying contents of sodium chloride, and contains not less than 12% K_2O.

Term	Terminology
Kaolinite	An aluminosilicate mineral of the 1:1 crystal lattice group: that is consisting of one silicon tetrahedral layer and one aluminium octahedral layer of about 0.7 nm thickness.
Kernel	Edible seed portion of rice, maize, sorghum etc.
Kharif crops	Crop grown during the main monsoon season (April to September), such as kharif sorghum, maize, pearl millet, rice, cotton etc.
Kieserite	Magnesium sulphate ($MgSO_4.H_2O$) containing 27% MgO and 22% S.
Kilowatt	A unit of power equal to 1,000 watts and it equals to energy consumption @ 1,000 joule/sec. It is usually used for electrical power.
Koppen's classification (climate)	The classification scheme which relates climate to vegetation while providing an objective numerical basis for defining climate types in terms of climatic elements.
Kotka phosphate	A product obtained by partial acidulation of ground rock phosphate with sulphuric acid and containing not less than 16% of available P_2O_5' of which at least 6% is water soluble.
Kreb's cycle	A cycle of reactions which plays a central role in the metabolism of many heterotrophic organisms, capable of respiratory mechanisms. It involves an enzymes system which converts pyruvic acid to CO_2 in the presence of O_2 with concomitant release of energy in the form of ATP molecules.
Krishi Vigyan Kendras (KVKs)	The schemes on KVK were launched by the ICAR during 1974 as an innovative institution for vocational training in agriculture and allied areas. Skill training is imparted to the trainees through the methods of teaching by doing and learning by doing. The KVK aims at practical training to practicing farmers, farm women as well as school dropouts who wish to acquire training for self-employment.
Kurtosis	The departure of a symmetrical frequency distribution from the normal by excess (platy kurtosis) or deficiency (lepto kurtosis) in its shoulders as opposed to tail and centre.

Term	Terminology
Label	All written printed or graphic matter on or attached to the economic poison, or the immediate container thereof, and the outside container or wrapper to the retail package of the economic poison.
Label (isotopic)	A labeled compound or molecule which contains one or more radioactive atoms (or stable atoms of different mass) as a part of its structure.
Labeled compound	A compound consisting of one or more of its components labeled with radioactive isotope.
Lab-to-land programme (LLP)	It is an instrument of transfer of technology. This programme was launched in 1979 as a part of the golden jubilee celebration of the ICAR. Under this project 50,000 small and marginal farm families and landless agricultural laboures were adopted by Research Institutes, Agricultural agencies and voluntary organizations with a view to guide, assist and help them to take advantage of the new technologies developed in the field of agriculture and improve the economic status of the farming lot.
Laminar flow	A flow in which fluid moves smoothly in streamlines in parallel layers or sheets (non-turbulent flow).
Land	That part of the surface of lithosphere which is not usually covered with water, forms the total natural and cultural environment with which production takes place.
Land (marginal)	Land from which it is doubtful whether a specified form of management can produce an income in excess of the costs of such management under given economic conditions, usually with reference to agricultural land.
Land (submarginal)	Land which is incapable of sustaining a certain use of ownership status without economic loss and/or land which cannot be profitably used for agriculture.
Land breeze	A local wind which blows from land to sea during the night.
Land capability	The suitability of land for use without damage. It is an expression of the effect of physical land conditions on the total suitability of land for different uses without damage for crops that require regular tillage, for grazing, for woodland for wildlife and other recreations.

Term	Terminology
Land capability classification	A grouping of the mapping units of a soil conservation survey into land capability units, subclasses, classes and general divisions.
Land capability map	A map showing land capability units subclasses, and classes, or a soil conservation survey map coloured to show land-capability classes.
Land characteristic	An attribute of land that can be measured or estimated and which can be employed as a means of describing land qualities or distinguishing between land units of differing suitabilities for use.
Land clearing	Removal of heavy vegetative growth from the land prior to land preparation.
Land drain	A drain for drawing of water from land.
Land drainage	The removal of free water both from the land surface and from the soil of the crop-root zone.
Land equivalent ratio (LER)	It denotes the relative land area under sole crop required to produce the same yield as obtained under a mixed or an intercropping system at the same management level. It is calculated as sum total of the ratios of yield of each component crop in an intercropping or a mixed cropping system to its corresponding yields when grown as a sole crop.
	Ratio of the area needed under sale cropping to one of intercropping at the same management level to give an equal amount to yield. LER is the sum the fractions of the yield of the intercrops, relative to their sale crop yields.
Land evaluation	The process of assessment of land performance when used for specified purposes, involving the execution and interpretation of surveys and studies of landforms, soil, vegetation, climate and other aspects of land in order to identify and make a comparison of promising kinds of land use in terms applicable to the aims of evaluations.
Land facet	A land unit (q.v.) with climate, landforms, soils and vegetation characteristics, which for most practical purposes may be considered as uniform. A subdivision of a land system.

Term	Terminology
Landforming	Tillage operation which move soil to create desired soil configurations. Forming may be done on a large scale, such as contouring or terracing or on a small scale, such as ridging or pitting.
Landgrading	Reshaping of the field surface to a desired elevation and slope. It is necessary in making a suitable field surface to control the flow of water, to check soil erosion and to provide surface drainage.
Land leveling	The reshaping of the land surface to facilitate a more uniform application of irrigation water or for preventing soil erosion.
Land planning	Tillage operation that cuts and moves small layers of soil to provide a smooth and refined surface.
Land reclamation	Making land capable of more intensive use by changing its character and/or environment through such operations as land clearing, controlling erosion, construction of irrigation reservoirs etc.
Landside	That parts of the plough those sides along the face of the furrow wall. It helps to slice on the ground and also helps in stabilizing the plough while it is in operation. Land side is fastened to the frog, another part of the plough with the help of bolts.
Land suitability	The fitness of a given type of land for a specified kind of land use.
Land suitability rating	The partial suitability of a land unit for a land utilization type, based on one land quality or a partial set of land qualities. Land suitability ratings are combined to give a land suitability class.
Land type	A union of locations within which values of cropping pattern determinants are the same.
Land-use patterns	Alternative ways to utilize available land resources over the time for agricultural production.
Land-use planning	The development of plans for the uses of land over long periods, will best serve the general welfare, together with tile formulation of ways and means for achieving such uses.

Term	Terminology
Land-utilization index	It is defined as the number of days during which the crops occupy the land during a year divided by 365 and can be expressed as a fraction or as a percentage.
Langlay	A unit for radiation measurement. It is also equal to g call cm².
Lapse rate	The rate at which temperature decrease in a rising and expanding air parcel.
Large plots	Plots from 1/200 to 1/100th hectare in size for the comparative testing of varieties with the use of farmers seeding and harvesting methods under ordinary farm conditions.
Latent heat	The heat released or absorbed per unit mass by a system during a change of phase. In meteorology, at 0°C the latent heat of vaporisation, fusion and sublimation of water are about 600, 80 and 680 call g, respectively.
Lateral movement	Movement of herbicide through soil, generally in a horizontal plane, from the original site of application.
Laterite soils	Soils formed in situ by the leaching out of the bases from the parent rock under monsoon conditions of alternate dry and wet seasons. These are characterized by poor fertility and deficient in N, P, K and Ca.
Laterization	Process of leaching out completely of Ca at or near the surface of the soil in heavy rainfall areas under tropical conditions, due to intense weathering taking place to a great depth. Fe and Al concentrate at or near the surface of the soil.
Lathyrism	Paralytic condition of animals caused by consumption of forage or seed of certain plants of genus *Lathyrus.*
Latin square	An experimental design in which there are same number of rows and columns as treatment, with each treatment occurring once and only once in every row and column, and is suitable when soil fertility gradient is in two directions.

Term	Terminology
Lattice design	Any experimental design in which each replication consists of several blocks with randomization occurring within these blocks making possible the comparison of a large number of treatments.
Lattice energy	The energy required to separate the ions of a crystal to an infinite distance from each other.
Lattice structure	The orderly arrangement of atoms in a crystalline material.
Law of additive probability	It states that if two events A and B are mutually exclusive (the probability of occurrence of either A or B is the sum of individual probability of A and B.
Law of average product	Average product is the total output divided by the total input.
Law of constant return	Constant productivity holds true if all units of the variable factor which are applied to the fixed factor, result in equal additions to the total output.
Law of diminishing return	Diminishing productivity of the variable factor exists when each additional units of input adds less to output than the previous unit.
Law of minimum	States that when a character such as yield is controlled by number of separate factors; the rate of the process is limited by the factor which is present in relative minimum.
Law of multiplicative probability	It states that if two events A and B are independent, the probability of their occurrence is same as the product of their individual probabilities.
Law of statistical regularity	This law states that a moderately large number of items chosen at random from among a very large group are almost sure on the average, to have the characteristics of the large group.
Lay-by	Refers to the stage of crop development (or the time) when the last regular cultivation is done.
Lay-by application	It is the application of herbicides with or after the last cultivation in crops, *e.g.,* after ridging in sugarcane and cotton.
Lay-out	The placement of experimental treatment on the experimental site where it may be over space, time or type of material.

Term	Terminology
LC_{50}	The concentration of a chemical in air or water that will kill 50 percent of the organisms in a specific test situation.
LD_{50} (pesticide)	Abbreviation of median lethal dose. It indicates the amount of toxicant necessary to effect a 50% kill of the pest being tested. Measured in mg/kg of body weight.
Leach	Usually refers to movement of water through a soil which may move soluble plant food or other chemicals.
Leaching	Is the process of removing soluble material by the passage of water through the soil?
	Refers to downward movement of water through a soil; this may involve moving soluble chemicals and plant foods along.
Leaching efficiency	The ratio of the average salt concentration in drainage water to an average salt concentration in the soil water of the root zone when near field capacity.
Leaching requirement	It is the fraction of water entering the soil that must pass through the root zone in order to prevent soil salinity from exceeding a specified value.
Leaf Blade	The expanded flat portion of a leaf.
Leaf-area duration (LAD)	It is a measure of the ability of the plant to produce and maintain leaf area and is obtained by integrating the leaf- area index over crop-growth period. It is usually expressed in days or weeks.
Leaf-area index (LAI)	The ratio between the area of the surface of green leaves and ground area covered.
Leaf-area ratio (LAR)	The ratio of leaf area to dry weight of plant.
Leeve	A bank of earth used to hold irrigation water within certain limits so that uniform irrigation of the entire field is obtained. An earthen dam placed at varying distances from the bank of a river to serve as barrier to adjacent low lying land during floods.

Term	Terminology
Legume	The word legume is derived from the Latin word legre (to gather) because the pods have to be gathered or picked by hand, as distinct from 'reaping' the cereal crops. The legumes belong to family Leguminoseae and form nitrogen-fixing nodules on their roots.
Legume inoculation	The addition of proper strain of nitrogen fixing bacteria to legume seed or to the soil in which the seed is to be planted.
Leibigs law of minimum	The growth and reproduction of an organism are determined by the nutrient substance that is available in minimum quantity, the limiting factor.
Lemma	The lower of the two bracts enclosing each floret in the grass spikelet.
Lenticel	An opening in the cork of roots and stem through which exchange of gases occurs.
Lento capillary point	The point when the capillary movement of water in a soil becomes sluggish and ineffective.
Lero morphic	Possessing structural features in plants which provide resistance to draught.
Lethal	So detrimental as to cause death.
Lethal gene	A gene that is capable of bringing, labour death. Lethal genes may be either dominant or recessive and kill the organism at any stage of its development.
Leucinization	The paling of soil horizons by disappearance of dark organic materials either through transformation to light coloured ones or through removal from the horizons.
Leucoplast	A colourless plastid in which starch is frequently formed.
Level terrace	A terrace that follows the absolute contour as contrasted with graded terrace.
Levelers	The two-wheeled automatic type leveler is usually used for the fine grading of small and medium-size fields.
Leveling	The tillage operation in which the soil is moved to established a desired soil-elevation slope.

Term	Terminology
Levels of inputs	Three levels of inputs are commonly recognized.
Ley crops	Any crop or combination of crops is grown for grazing or harvesting for immediate or future feeding to livestock.
Ley farming	A rotation of arable crops requiring annual cultivation and artificial pastures occupying field for two years or longer.
Life-saving irrigation	The irrigation given to a dry land or rainfed crop to save it from total failure due to long dry spells or severe soil-moisture stress, the source of irrigation generally being harvested water.
Light	The visible portion of the spectrum. The term IS sometimes incorrectly used to include the ultraviolet and infrared portions as well.
Light compensation point	The light intensity at which photosynthesis and the respiration rates in a leaf or other chlorophyllous organ are equal. At this light intensity the volume of CO_2 being released in respiration is equal to the volume of CO_2 being consumed in photosynthesis, while the opposite is true for oxygen.
Light quality	Composition of light in terms of wavelength of monochromatic radiation.
Light reaction	The photolysis of water, the photochemical reduction of pyridine rlucleotide and photophosphorylation are together called as light reaction of photosynthesis.
Light saturation intensity	The point in light intensity at which an increase in it fails to speed up photosynthesis.
Light soil	A soil that is easy to work implements in or easy to cultivate as it lacks cohesiveness, such as sandy and sandy-loam soils.
Light transmission ratio	Light intensity at the ground level of crop/plant population divided by the light intensity at the top of the crop canopy/population.
Lignin	An organic chemical which occurs in the walls of cells, especially wood cells and makes substance resistant to decomposition.

Term	Terminology
Lime	Generally the term lime or agricultural lime is applied to ground limestone, hydrated lime or burned lime which is used as amendments to reduce the acidity of soils. In strict chemical terminology lime refers to calcium oxide (CaO).
Lime chlorosis	A yellowing off foliage commonly occurring in soils high in lime or irrigated with water high in lime. It is caused by a deficiency of available iron, zinc or copper because insoluble calcium compounds are readily formed with these minerals.
Lime requirement	The mass of agricultural limestone or the equivalent of other specified liming material required per acre to a soil depth of 6 inches (or 2 million pounds of soil) to raise the pH of the soil to a desired value under field conditions.
Limestone	The term refers to rocks consisting chiefly of calcium carbonate or calcium and magnesium carbonates. When magnesium carbonate is present in a large percentage, limestone is called dolomite limestone.
Liming	The application of lime to land primarily to reduce soil acidity and to supply calcium for plant growth.
Liming material	Product containing one or both of the elements calcium and magnesium, generally in the form of an oxide, hydroxide or carbonate, principally intended to maintain or raise the pH of soil.
Limiting factor	A factor which is in short supply or exceeds the optimum limits for an organism. A limiting factor may operate both when deficient or in excess.
Linear response	When the output (Y) goes on changing proportionately along with input (X) the response is said to be linear.
Linkage	The tendency of certain genes to remain together because they are situated on the same chromosome.
Lint index (LI)	The weight of lint per seed cotton or 100 seeds.
Lint length (cotton)	The length of unicellular spindle hairs arising from the seed-coat collectively known as lint, the length varies from less than 1/8 of an inch to 2.25 inches (0.32-5.72 cm).

Term	Terminology
Lipids	Compounds insoluble in. water but soluble in organic solvents, chemically either esters of fatty acids or hydrolytic products of such esters, and can be used in the metabolism of living organisms, *e.g.,* fats, oils, waxes etc.
Lipophilic	The character of substances having greater affinity for oil and other non-polar solvents.
Lipoproteins	Lipids in combination with proteins.
Liquid fertilizer distributor	A machine for applying mineral fertilizer in liquid form which either spreads the fertilizer on the surface of the ground or injects it into the soil.
Liquid fertilizers	Commercial fertilizers in liquid form. Such fertilizers are chiefly anhydrous ammonia, aqueous solution of nitrogen, and some mixed fertilizers. Liquid fertilizers are applied to the soil through irrigation water as starter solutions, and with the help of special equipment.
Liquid manure	Liquid resulting from animal urine and litter juices or from a dung heap.
Liquid scintillation (nuclear)	A method of counting beta particles, particularly of low energy, by mixing the sample with an organic solvent containing a scintillator. The light flashes emitted are registered using photomultiplier tube detection.
Lister	An implement for furrowing land, often having a planting attachment.
Listing	A tillage and land forming operation using a tool which splits the soil and turns two furrows laterally in opposite directions, thereby providing a ridge-and-furrow soil configuration.
Litre	The French standard measure of capacity, equal to 61.028 cubic inches, or consisting 1,000 milliliters (ml), used for measuring liquid substance or solution etc. The English Imperial gallon is being 4.5 litres.
Lithophytes	Plants which are usually found on rock surfaces.
Litmus	A colouring matter produced from certain lichens. Red and blue litmus papers are used for chemical tests.

Term	Terminology
Litter	A bed in which straw is used as a bed for animals, the young produced at a birth by a quadruped.
Littering	Accumulation of organic matter arid associated humus on the mineral soil surface to a depth of less than 30 cm.
Live storage	The water stored above the outlet in a reservoir. Loam: Soil material that contains 7-27% clay, 28-50% silt and less than 52% sand.
Loamy	A broad grouping of soil texture classes includes sandy loams, clay loams, loamy silt and silt loam texture.
Local control	Making use of greater homogeneity of adjacent areas in a field, the entire experimental units are divided into compact blocks which are known as local control. Local control will reduce the experimental error thereby increasing the precision of the experiment.
Local variety	A mixture of different types and is well adapted to local environment. It is endemic to an area with its origin dating back to several hundred years.
Localized placement of fertilizers	This refers to the application of fertilizer into the soil close to the seed or plant. This method is specifically adopted with phosphatic fertilizers for three reasons, *viz.* restricted contact of fertilizers with soil lessens fixation of phosphate, necessary plant food is placed within easy reach of plant roots, and application of fertilizers in a band along the rows does not readily furnish nutrients to weeds growing between the rows.
Lodging (crop lodging)	A state of standing crop in the field being flattered before harvest by heavy rain and/or wind either as a result of excessive soil fertility, particularly levels of available nitrogen or the presence of fungal disease on the stem.
Loess	Geological deposit of relatively uniform fine material, mostly silt, presumably transported by wind.

Term	Terminology
Logarithmic sprayer	A sprayer devised for applying a chemical at a steadily decreasing rate.
Logarithmic transformation	The logarithmic transformation ($Y = \log x$) is useful when dealing with material on which the effect of other variables may be expected to be proportional t numbers involved. For example, data relating to the influence a given increase in temperature on the number of insect population.
Long-day plants	Plants in which flowering is initiated or hastened under day-length above a definite value or when the photoperiod is greater than a certain critical minimum, *e.g.,* wheat, barley, oat etc.
Long-wave radiation	Electro-magnetic radiation with a wavelength greater than 0.8 microns.
Lopping	Pruning or cutting away aerial parts of shrubs or trees.
Low inputs	No significant use of purchased inputs such as artificial fertilizers, improved seeds, pesticides or machinery. Traditional farming in developing countries.
Low-analysis fertilizer materials	The fertilizers containing a low percentage of plant nutrients, usually less than 30%.
Lowland rice	Rice grown after wetland preparation of field (based on land and water management) or rice grown with 5-50 cm standing water (based on water regime).
Low-volume spraying	Spraying, when spray material deposited in small drops do not form pools and obtained by spraying liquid up to 200 litres/ha.
Lumen	Measure of illumination, 1 lumen sq ft = 1 ft candle.
Lux	A unit of measure of illumination (solar radiation), 1 lumen sq ft = 10.764 lux.
Luxury consumption (nutrients) uptake	The absorption of nutrients by plants in excess of their need for growth. Luxury concentrations during early growth may be utilized in later growth.
L-value (tracer studies)	It is a measure of the isotopically exchangeable fraction of phosphate in soil and is used to determine the quantity of soil phosphate in labile form.

Term	Terminology
Lysimeter	A device for measuring percolation and leaching losses from a column of soil under controlled conditions.
Macronutrient	A chemical element necessary in relatively large amounts (usually 500 parts per million in the plant) for the growth of plants. These elements consists of C, H, O, Ca, Mg, K, P, S and N.
Macro pores	Soil characterizing size of pores having diameter of 100 microns or more which maintains high air circulation and increase draining of soil water and nutrients.
Magnesia (magnesium oxide)	A product consisting chiefly of the oxide of magnesium.
Magnesium sulphate	Usually the product known as kieserite ($MgSO_4$. H_2O). The highly hydrated form Epsom salts ($MgSO_4$. $7H_2O$) is suitable for foliar application.
Maize sheller	An equipment to remove maize grain from cobs.
Major project (irrigation)	A project which can irrigate more than 10,000 ha.
Male sterile	Complete or partial failure of plant to produce mature reproductive pollen cells.
Malting	The process of germination of grain generally- barley to develop the enzyme diastase.
Manday of labour	The amount of work expected from a farm worker in an 8-hour day, operating at the average rate of performance using adequate equipment.
Manganese toxicity	Symptoms are brown spots on older leaves, leaf-tip drying and high sterility. It is a disorder of dryland rice on acid soils.
Manometer	An instrument for measuring pressure usually consists of a V-shaped tube containing liquid which moves proportionately with changes in pressure upon the liquid.
Manual	Pertaining to or performed with the hand.
Manure	The excreta of animals (dung and urine) with straw or other materials, used as the absorbent. The decomposed manure is called farmyard manure or farm manure or barnyard manure. The average composition of well-rotted farmyard manure is 0.5% nitrogen, 0.3% P_2O_5 and 0.5% K_2O.

Term	Terminology
Manure application	Process of spreading manure on the soil and in some cases mixing it with soil.
Marginal benefit: cost ratio	It is a good index of the economic advantage of a new pattern. It helps to evaluate a series of alternatives to the existing pattern and to select the most remunerative one to the farmer.
Marginal product (MP)	Marginal product is the ratio of change in total output as related to the change in input.
Marl	Soft and unconsolidated deposits of calcium carbonate usually mixed with varying amounts of clay or other impurities.
Marsh	A periodically wet or continually flooded area with the surface not deeply submerged. Covered dominantly with sedges, rushes or other hydrophytic plants.
Mass flow	A liquid exposed to a greater pressure in one region moves from the region of higher to that of lower pressure and this phenomenon is called mass flow.
Mass number	The number, indicated by the symbol 'A', which represents the total number of nucleons in a nucleus.
Mass selection	The selection of a number of plant heads or seeds, phenotypically superior in characters from the field populations, harvesting and bulking their produce together for sowing the next year's crop, and repeating this process till the desired improvement is achieved.
Mass-spectrometer	An instrument that separates (ionized) atoms or molecules according to mass and produces a spectrum indicating relative intensity as a function of mass.
Matric potential	It may be defined as the amount of work that a unit quantity of water in an equilibrium soil water (or plant water) system is capable of doing when it moves to another equilibrium system identical in all respects except that there is no matrix present.
Maturity	The stage at which a plant is capable of reproducing itself by seed.

Term	Terminology
Maturity or mature grain stage	When grain colour in the panicle begins to change from green to yellow and 90-100% grains are fully developed, hard and free from green tint.
Maximum available water	The quantity of water represented as the difference between field capacity and wilting coefficient. This is the moisture that can be readily extracted by the plant and used for normal growth.
Maximum cropping	The highest possible production per unit area per unit time without considering cost of production or net return.
	Attainment of highest possible production per unit area per unit time without regard to cost or net return.
Maximum residue limit (MRL)	Means the maximum concentration of a residue that is legally permitted or recognized as acceptable in or on a food, agricultural commodity or animal feedstuff.
Maximum thermometer	A thermometer used for measuring the highest temperature attained during a day. It is usually mercury in glass thermometer.
Maximum tillering stage	The stage of the crop at which it produces maximum potential number of tillers.
Meadow	Area covered with grasses and/or succulent forage legumes grown primarily for hay.
Mean	The mean is the arithmetic average and is obtained when the sum of the values of individuals in the data is divided by the number of individuals in the data
Mean deviation	Sum of the deviation of each score from the mean, without regard to sign divided by number of scores.
Mean life (nuclear)	The average life of a radioactive atom and is equal to the reciprocal of the decay constant.
Mean sea-level	The mean plane at which the tide oscillates; the average height of the sea for all stages of the tide. At any particular place, it is derived by averaging the hourly tide heights over a 19-year period.

Term	Terminology
Mean square	The variance so called because it is the average of the squared deviations from the arithmetic mean of a series of values.
Mechanical analysis	The determination of the relative distribution of the size group of ultimate soil particles. The important steps in mechanical analysis are (i) separation of all particles from each other so as to cause complete dispersion into ultimate particle, and (ii) measuring the amounts of each size group in the sample.
Mechanical impendence	The mechanical resistance of the soil to the movement of plant roots or tillage tools.
Mechanical stability	The degree of resistance of the soil to the deformation or break down by a specified mechanical force.
Mechanical strength	The mechanical resistance of soil to deformation or separation by forces.
Mechanism of action	Precise biochemical or biophysical reaction or series of reactions that culminate into the final or ultimate effect of a herbicide; many herbicides have a primary and' secondary mechanism of action.
Mechanized farming	It is the farming in which machine-drawn implements are used for the reduction of labour requirement or elimination of manual work, timeliness of operations and improved quality of husbandry, resulting in higher output and better quality of produce for increased profit.
Median	The median is the values which are located in the middle of a series when the observations are arranged in order of magnitude and it divides the series into two equal halves, half the number of observations lying above it and half below.
Medium (growth)	Any material which provides nutrients for the growth and multiplication of micro-organisms or plant seedlings. A defined medium is one in which all the constituents are quantitatively known.
Medium black soil	These are highly agrillaceous (30-50% clay) and crack deep in summer, contain high amount of Fe, Ca, Mg, K, lime and poor in organic matter, N, P and drainage.

Term	Terminology
Medium project	An irrigation project irrigating an area between 2,000 to 10,000 ha.
Meiosis	The cell-division which halve the number of chromosomes from the diploid to the haploid number; also known as reduction division.
Mellow soil	A soil that is easily worked or penetrated.
Melting point	A characteristic temperature at which the solid state of a substance is in equilibrium with its liquid state under a pressure of one atmosphere.
Meristem	A tissue at the stem or root tips wherein active cell division takes place.
	A tissue primarily concerned with protoplasmic synthesis and formation of new cells by division.
Mesocotyl	The elongated portion of the axis between the cotyledon and the coleoptiles of a grass seedling.
Meson	A particle with a mass greater than that of the electron but less than that of the proton.
Mesophytes	Plants with characteristics intermediate between xerophytes and hydrophytes and are adapted where moisture and aeration conditions lie well between both extremes.
Mesopores	Soil characterizing. pores size having diameter ranging from 30 to 100 microns which helps in capillary water and nutrient movement or increase conductivity of water, aeration and drainage.
Metabolic body size	Weight of the animal rose to the three-fourths power. The ration of the animals is calculated and standardized on the basis of metabolic body size.
Metabolism	The aggregate of all physical and chemical processes constantly taking place in living organisms, including those which use energy to build up assimilated materials (anabolism) and those which release energy by breaking them down (catabolism).
Metabolite	A compound derived from metabolic transformation of an herbicide by plants or other organisms.

Term	Terminology
Metabolizable energy (ME)	Food intake gross energy-fecal energy-energy in the gaseous products of digestion-urinary energy.
Metal dam	It is a water-control device, made up of metal sheet with sharp edge at the bottom fitted with a handle which is portable and of desired size used for obstructing flow of irrigation water in canal or channel.
Metamorphic rocks	Formed from pre-existing rocks through action of heat and pressure or due to other geological processes originating within the earth.
Methaemoglobinaemia	A condition associated with nitrate toxicity as a result of consuming nitrate rich waters. Nitrate transforms ordinary oxyhaemoglobin into inactive methaemoglobin and infants suffering from this condition have been referred .to as blue babies due to development of grayish-blue skin and the problem as blue baby syndrome.
Method demonstration	A type of demonstration in which the extension worker demonstrates an improved technology or farm practice for the benefit of the farming community.
Micro plot	One of many similar small plots used in determining accurately the comparative performance of varieties in the field.
Micro relief	Small scale, local differences in topography including mounds, swales, or pits that are only a few feet in diameter and with elevation differences of upto 6 feet.
Microbial degradation	It is breakdown or detoxification and inactivation of the soil applied herbicides by soil micro-organisms.
Microbial insecticide	Use of pathogens for killing insects and is free from harmful residue, non-phytotoxic to crops at higher doses and harmless to beneficial insects.
Microclimate	It is the local climatic condition near the ground or area around plants (up to 2 m height), resulting from the modifications of the general climatic conditions by local difference in relief, exposure and cover etc.

Term	Terminology
Micro-environment	The environment close enough to the surface of a living or non-living object to be influenced by it.
Micrometeorology	Study of the meteorological conditions in the immediate surroundings of an organism.
Micron	One millionth of a meter (10^{-6} m) or one thousandth of a millimeter (10^{-3} mm).
Micronutrient (minor element or trace element)	A chemical element necessary only in extremely small amounts (usually less than 50 parts per million in the plant) for the growth of plants. These elements include B, Cl, Cu, Fe, Mn, Mo and Zn.
Micro-organism	A micro scopic form of life.
Micro pores	Soil characterizing pores size having diameter ranging from 3 to 30 microns which plays an important role in water and nutrient retentions with slow capillary flow.
Mid-day depression or mid-day deficits	Even during periods of high radiation and low humidity those plants growing in soils nearing field capacity may be subjected to severe water stress. The phenomenon is known as mid-day depression or water deficits.
Middle breaking	The use of lister in a manner that opens the furrow midway between two previous rows of plants.
Middle buster	A double shovel plough of lister.
Mid-season correction	It is a contingent management practice to overcome unexpected and/or unfavorable weather conditions, *e.g.,* thinning or population adjustment according to the availability of moisture in the soil during the dry spell.
Milk grain stage	The stage where the contents of the caryopsis (the starch portion of the grain) are first watery but later turn into milky white liquid which can be squeezed out.
Milli equivalent per liter (me/l)	A milliequivalent of an ion or a compound in one litre of water.
Millibar	Pressure unit used in meteorology or agricultural meteorology. 1 millibar (mb) = 10^{-3} bar = 0.750062 mm of mercury.

Term	Terminology
Milling recovery	Weight of commercial rice recovered from the original rough rice (paddy), usually expressed in per cent.
Mineral soil	A soil consisting predominantly of and having its properties determined predominantly by mineral matter. Usually contains 20% organic matter but may contain an organic surface layer up to 30 cm thick.
Mineralization	The conversion of an element from an organic form (immobilized) to an inorganic state (available) as a result of microbial decomposition.
Minimum thermometer	Thermometer used for measuring the lowest temperature attained during a day. It is generally a alcohol or spirit-in-glass thermometer.
Minimum tillage	The soil manipulation necessary to meet the minimum tillage requirements for crop production.
Minor project	An irrigation project with a capacity to irrigate less than 2,000 ha.
Miscible	Two or more liquids which, when combined together, form a uniform stable mixture.
Miscible liquids	Two or more liquids capable of being mixed and which will remain mixed under normal conditions.
Mistscherlich's law	It assumes that a plant produces its maximum yield if all conditions are ideal; if any essential factor is absent, there is corresponding reduction in the yield and assuming that an increase in yield per unit increment of the lacking factor is proportional to the decrement from the maximum.
Mitosis	The division of a nucleus of somatic cell involving the longitudinal splitting of the chromosomes and the regular distribution of chromosome halves to two daughter nuclei. Mitosis insures that each of the two resulting nuclei will be exactly like the parent nuclei.
Mixed cropping	Growing of two or more crops simultaneously on the same piece of land, without any definite row arrangement.

Term	Terminology
Mixed farming	Farming system which involves the raising of crops and rearing of animals and for poultry. Mixed farming is based upon the principle that land should support animal and animal should support land. Mixed farming is a type of farming under which crop production is combined with livestock raising. The livestock enterprise is complementary to crop production programmes so as to provide a balanced and productive system of farming. Cropping systems which involve the rising of crops, animals and/or trees.
Mixed farming systems	Farming systems with integrated crops, livestock, and other possible household enterprises.
Mixed fertilizers	Mixed fertilizers consist of individual or straight fertilizer materials blended together to permit application in the field in one operation. Mixed fertilizers supply two or three major plant nutrients. The percentages of these nutrients are expressed as fertilizer grade, like 10-5-5.
Mixed grazing	Two or more classes of livestock, such as sheep and cattle, are grazed on the same pasture.
Mixed intercropping	Growing two or more crops simultaneously with no distinct row arrangement.
Mixed red and black soil	Light textured, free of carbonate, sandy clay, differs in depth, deficient in nitrogen, phosphate, organic matter and lime, medium in fertility on which variety of crops can be grown.
Mixing ratio	The ratio of the mass mv of water vapour to the mass ma of dry air with which the water is associated, mv/ma.
Mobile	Movement of element from one place to other parts, such as essential elements, nutrients, water etc. Mobility varies with the element.
Mobilization	It is a physiological process in which movement of elements, food or other materials occurs from one plant part to another part.

Term	Terminology
Mode	The mode is that value of the variate which occurs most frequently. In a frequency table the modal class is the class· which has the greatest frequency.
Mode of Action	The sequence of events that occur from an herbicide's first contact with a plant until its final effect (often plant death) is expressed.
Mode of action (mechanism of action)	This indicates how the herbicide kills or inhibits growth of plants.
Moderator (nuclear)	A material used to slow neutrons in a reactor. Neutrons are slowed down when they collide with atoms of light elements such as hydrogen and carbon-2 common moderators.
Moisture (soil)	Soil particle holds water on its surface in the form of a thin film with a certain force. Soil pore space contains both air and water in varying proportions. Soil moisture expressed in percentage on oven-dry basis either on weight basis or on volume basis. The terms soil moisture and soil water are used generally as synonyms.
Moisture content (seed)	Content of water in the grain on wet basis, expressed as percentage by mass.
Moisture content (soil)	The quantity of water present in soil usually expressed in percentage by weight, on oven-dry basis.
Moisture equivalent	The moisture retained in air-dried, screened sample of soil which has been wetted and drained in a standard manner and centrifuged for 30 min in a centrifugal field equal to 1,000 times gravity. Moisture equivalent is expressed as moisture percentage on a dry-weight basis and approximates field capacity for much medium and fine-textured soil.
Moisture percentage	It may be defined on the basis of dry weight as the weight of water per 100 units of weight of material dried to constant weight at a standard temperature or on basis of depth as the equivalent depth of free water per 100 units of depth of soil Numerically this value approximates the volume of water per 100 units of volume of soil.

Term	Terminology
Moisture regimes	The different levels of the available soil moisture (ASM as %), a testing factor or unit inputs used in conducting irrigation experiments, such as ASM at 100, 80 60, 40% etc.
Moisture stress	The tension at which water is held by the soil.
Moisture tension	The equivalent negative pressure to which water must be subjected in order to be in hydraulic equilibrium through a porous permeable wall or membrane, with the water in the soil.
Moisture-availability index	It is the ratio of assured rainfall (weekly or monthly) at 50% probability level to potential evapotranspiration of the corresponding period.
Moisture-deficit index (MDI)	This is the measure to estimate dryness of a region in percentage.
Moisture-release curve or characteristic curve	It is the functional relationship between soil-moisture tension and soil moisture content of a range from field capacity to wilting point.
Moisture-retentive soil	A soil that retains a high percentage of water, directly depending upon clay colloid content. Clay soils are more moisture retentive than sandy soils.
Molar concentration	The number of gram molecules of the solute present in 1,000 g of solvent, usually water.
Molar solution	A solution in which one g molecular weight of the substance is dissolved in the solvent and made up to one litre.
Molasses	Byproduct of sugar industry, viscous liquid left behind after sucrose is extracted, separated by centrifuging sugar in sugar factories, used in preparation of methylated spirit, power alcohol, or alcoholic beverages, yeast, cattle feed etc.
Mole	A unit of mass numerically equal to the molecular weight of the substance.
Mole drains	Cylindrical channels artificially formed in the subsoil using a mole plough without digging a trench from the surface.
Molecular biology	A field of biology which emphasizes the interaction of biochemistry and genetics in the life of an organism.

Term	Terminology
Molecular weight	The molecular weight of a substance is defined as the average mass of its molecule compared with the mass of an oxygen atom taken as 16.
Molecule	The smallest unit of a substance that can independently exist and retain its properties.
Mollisols	Soils with nearly black organic rich surface horizons and high supply of bases. They have mollic epipedons and base saturation greater than 50% in any cambic or argillic horizon. They lack the characteristics of Vertisols and must not have oxic or spodic horizons.
Monetary advantage index (MAI)	While assessing yield advantages of different intercropping systems, an index was formed by taking into consideration the relative money value of the produce.
Monetary index	A single productivity index which incorporates monetary value of all, crops in a system.
Monitoring study	Making systematic observations through well-designed procedures on a crop, cropping pattern, farm or experimental trials to relate resultant effects with observed factors or causes.
Monochromatic radiation	Electromagnetic radiation of a single wavelength. That is, all photons have the same energy.
Monocotyledon (Monocot)	Any seed plant having a single cotyledon or seed leaf. A plant having one cotyledon as in grasses.
Monocrop (mono culture)	A cropping system in which the same crop is grown year after year in the same field.
Monoculture	Repetitive growing of the same sale crop in the same land. Sole cropping. One crop variety grown alone in. pure stand at normal density.
Monoecious	A trait of having separate male and female flowers on the same plant.
Monohybrid	An F_1 that is heterozygous at one locus.
Monopolistic polyculture	The interaction between crop species has a positive net effect on one species and the negative effect on the other species, when grown together.

Term	Terminology
Monosaccharide	A simple sugar which cannot be split further into smaller sugar molecules, *e.g.,* pentoses and hexoses.
Monsoon	The seasonal wind of the Indian Ocean and Southern Asia, blowing from the south-west in summer and from the north-east in winter. Monsoon blows, commonly marked by heavy rains.
Montmorillonite (2:1 expanding type clay)	Two tetrahydral silica sheets having one octahedral alumina sheet sand witched between them. Layer thickness varies between 1.0 and 1.8 nm.
Mor	Humus which because of its poverty in bases has a very acidic reaction.
Morphological adaptation	Plants may have morphological adaptation such as growth habit, strength of stalk etc.
Morphology	The study of form, structure and development of an organism or plant.
Mound culture	Soil is separated into mounds varying in height (30–50 cm) and shape to fit local needs. Distance between mounds are determined by soil depth; amount of soil needed to construct mounds of a given height and type of crop to be grown.
Mower	A machine mainly used for harvesting grasses and forage crops.
Muck soil	Fairly well-decomposed organic soil material relatively high in mineral content, dark in colour, and accumulated under conditions of imperfect drainage.
Mulch	Any material such as straw, plant residues, leaves, loose soil or plastic film placed on the soil surface to reduce evaporation, erosion or to protect plant roots from extremely low or high temperature.
Mulch crops	To conserve soil moisture, such crops are grown.
Mulch farming	A system off arming in which the organic residues are not ploughed into the ground but are left on the surface.

Term	Terminology
Mulch tillage	Preparation of soil in such a way that plant residues or other mulching materials are specially left on or near the surface.
Mull	Humus which contains appreciable amounts of mineral bases.
Multidisciplinary approach	An approach in which several disciplines become involved in a project or programme with common general objectives.
Multiennials	The plant species which flower at an interval of a few too many years, after which they die off, *e.g.,* bamboos.
Multilocation testing	The testing of cropping-system technologies or new varieties generated at the research stations on other locations within the large target area to delineate the extra-zone as well as to finally verify technology performance before wide scale diffusion.
Multiple correlation	Multiple correlation coefficients measure the collective association of several variables with one another.
Multiple cropping	Growing two or more crops consecutively on the same field in the same year. In essence, represents a philosophy of maximizing crop production per unit area of land within a calendar year with minimum deterioration of soil.
	Intensification of cropping in time and space dimensions. Growing two or more crops in the same field in a year.
Multiple cropping index	Measures the sum of area planted to different crops and harvested in single year divided by the total cultivated area times 100.
Multiple cross	Method followed in the production of hybrid seed. A combination of more than 4 inbred lines made to obtain greater variability in the combination to suit varying agro climatic conditions.
Multiple resistance	When for same herbicide, two or more resistance mechanism in plant or weed species' occurs, it referred as multiple resistance.

Term	Terminology
Multiple-nutrient materials	The fertilizers containing more than one nutrient element. When fertilizers supply two major plant nutrients they are called binary fertilizers, *e.g.,* ammonium phosphate and nitrophosphate. Fertilizers supplying three major plant nutrients are termed ternary fertilizers, *e.g.,* ammonium potassium phosphate.
Multipurpose project	It is the construction of an irrigation project to serve more than one objective such as flood control, irrigation, power generation, navigation etc.
Multistorey or multitier cropping	It is a system of growing crops together of different heights at the same time on the same piece of land and thus using land, water and space most efficiently and economically, *i.e.,* coconut + pepper + pineapple + grass.
Multistoreyed	Growing plants of different heights in the same field at the same time is termed as multistoreyed cropping, *e.g.,* Eucalyptus, Papaya, and Berseem grown together.
Muriate of potash (KCl)	Commercial potassium chloride salt containing 60% ~O and is readily available to the plants.
Mutagen	Any substance, chemical or physical that induces mutations in organisms.
Mutagenic	Capable of causing genetic changes.
Mutant	Any organism, population or organisms, chromosome gene *etc.* which as a consequence of one or more mutation differs from the original type. The changed condition is inherited by offspring cells.
Mutation	Sudden, heritable variation in an animal resulting from an abrupt change in the genotype nature.
Mutation breeding	The creation of mutations at will and their utilization for crop improvement and production of new superior crop varieties.
Mutual co-operation	The yield of each species is greater than expected under intercropping situation.
Mutual inhibition	The competitive effects under intercropping situation are called mutual inhibition when the actual yield of each species is less than expected.

Term	Terminology
Mycorrhiza	The morphological association, usually symbiotic of fungi with the roots of seed plants. The association is referred to as ectotrophic, the fungal hyphae occur on the root surface and penetrate only the intercellular spaces and as endotrophic when the hyphae occurs mainly within the cells of the host plant.
Nastic movements	Movements which show no definite directional relation to the stimulus.
National Agricultural Research Project (NARP)	It is a project launched by the ICAR to strengthen the regional research capabilities of the agricultural universities in India, with an objective of conducting need based, location-specific and production-oriented research in all the agro-climatic zones identified in the area of its jurisdiction.
National demonstrations	The national demonstration project was launched by the ICAR in 1965 as an instrument of transfer of technology. The main objectives are (i) demonstrating to the farmers the production potential of land by adopting high yielding varieties, multiple-cropping programmes and other scientific package of practices, (ii) providing research workers a first-hand information on problems faced by farmers and factors limiting crop yields.
Native pasture	A pasture covered with native plants or naturalized exotic plants.
Natural erosion	It refers to erosion when the land is in its perfect equilibrium and undisturbed environmental condition under the cover of vegetation.
Natural farming	It is a system of alternative agriculture in which the plants are grown as natural entities without manipulation of soil. It is a system of farming which uses no machines, no prepared fertilizers and no chemical but yields normal harvests.
Natural radioactivity	The radioactivity which is observed in unstable isotope that exists in nature.
Natural selection	An important feature of certain theories of evolution, according to which agents other than man determine which members of a population, will survive.

Term	Terminology
Ndff (nitrogen derived from fertilizer)%	Percentage of N in the plant derived from the fertilizer applied.
Necrosis	Local death of cells due to disintegration of the cellular structure with destruction of the nucleus and liquification of the cytoplasm.
	Localized death of tissue usually characterized by browning and desiccation.
	The death of plant tissue.
Negative allelopathy	Some crops may stimulate growth of the associated crop by release of hormone-like substances; contrary to allelopathy.
Net area sown	Class of land under revenue classification represents actual extent of land sown to crops during a given year, when cropped more than once in any year, its area is added to the net area sown to arrive at the total area sown during the year.
Net assimilation rate (NAR)	It is increase in dry weight of plant per unit leaf area per unit time.
Net duty of water	The amount of irrigation water delivered to the farm for raising a crop; measured at the point of delivery.
Net energy (NE)	Difference between metabolizable energy and heat increment; includes the amount of energy used either for maintenance only or for maintenance plus production.
Net irrigation requirement (NIR)	The depth of irrigation water exclusive of precipitation, carry-over soil water or groundwater contribution or other gains in the soil moisture, that is required for plant growth; the amount of irrigation water required bringing the soil-moisture level in the effective root zone to the field capacity.
Net photosynthesis	When photosynthesis is measured with correction for respiration.
Net plot	The area from which yield or other characters is measured. It is also known as the net area of the plot.
Net productivity	The arithmetic difference between calories produced in photosynthesis and calories lost in respiration.

Term	Terminology
Net radiation	The difference between the radiation absorbed by an object and heat emitted.
Net return	It is the income obtained or remaining after deducting cost of cultivation from the gross return.
Neutral detergent fibre (NDF)	Lignin + cellulose + hemicelluloses; Hemicellulose = NDF + ADF.
Neutral soil	A soil that is neither significantly acidic nor alkaline, having pH between 6.6 and 7.3.
Neutron	A neutral elementary particle having a mass number of 1. In the free state (outside the nucleus) it is W\Stable, having a half-life of about 12 min.
Neutron moisture meter (depth probe)	Neutron source and detector in a cylindrical container that can be lowered down a tube in soil or other material and calibrated to indicate moisture content.
Neutron scattering	A means for determining the moisture content of the soil. Fast neutrons loose energy and are slowed down when they collide with hydrogen atoms. The slow neutrons are caught by a detector and counted by a counting unit.
Neutrophiles	The plants which can grow in neutral soils (pH range 6.5–7.4).
Nick	In hybrid seed production, the parents are said to nick when they produce higher yield as a result of synchronized flowering.
Nicotine	An alkaloid produced as a secondary plant product; by trichomes (glandular hairs present on the leaf surface) as secretion in tobacco, which has some insecticidal properties, especially against aphid.
Night soil	Night soil is human excrement, solid and liquid. In India, it is directly applied to the soil to a limited extent and as town compost to a large extent. In cities which have sewage facilities, sewage water, and sludge are used indirectly to raise crops. On an average night-soil contains 5.5% nitrogen, 4.0% phosphorus (PPs) and 2.0% potash (K_2O) on oven-dry basis.

Term	*Terminology*
Nitrate of soda and potash	Chiefly the sodium and potassium salt of nitric acid containing not less than 15 % of nitrate nitrogen and 10% of potash (as K_2O).
Nitrate reductase (NR)	A metalloflavoprotein enzyme that catalyzes the reduction of nitrate to nitrite.
Nitrate reduction	Reduction of nitrate to the level of ammonia with the help of enzymes.
Nitrate toxicity	Nitrate accumulation (0.43% as NO_3) in forage can cause acute poisoning in the livestock. Although nitrate itself is relatively non-toxic, it may be reduced to nitrite by soil microorganisms in cut forage or hay stored under moist conditions. Rumen micro-organisms also readily reduce nitrate to nitrite which is then absorbed into the blood stream where it converts haemoglobin to methaemoglobin due to the oxidation of Fe^{+2} to Fe^{+3} by nitrite. The methaemo-globin prevents the transport and release of oxygen by the blood and conversion of a substantial propor-tion of haemoglobin, therefore, results in internal asphyxiation.
Nitrate-nitrogen	Presence of nitrogen in form of nitrate denoted NO_3-N. It occurs in soil medium as the ammonia is converted into nitrate due to microbial activity *(Nitrosomonas* and *Nitrobacter)*. Nitrate fertilizers are carriers of NO_3-N. Most of plant species use NO_3-N.
Nitrification	It is a 2-step process in which the ammonia or ammonium is converted to nitrite and finally to nitrate and these 2 steps are carried out by the autotrophic micro-organisms *Nitrosomonas* and *Nitrobacter,* respectively.
Nitrification inhibitors	The chemicals used to control nitrogen transfor-mations in the soils. Due to their toxic effect on nitrifying bacteria, they temporarily inhibit the production of nitrate, *e.g.,* N-Serve, nitrapyrin, potassium azide etc. Application rate 0.2-0.6 kg of active ingredient.
Nitrogen	Colourless, tasteless, odourless gas, which occurs in the atmosphere to the extent of 78% by volume. An essential plant nutrient, taken up by plants in the form of ammonium or nitrate ion.

Term	Terminology
Nitrogen assimilation	The incorporation of nitrogen compounds into cell substances by living organisms.
Nitrogen cycle	The sequence of transformation undergone by nitrogen, wherein it is used by one organism, later liberated upon the death and decomposition of the organism, and is converted by biological means to its original state of oxidation to be reused by another organism.
Nitrogen deficiency	Deficiency during early growth stages results in yellow to yellowish-green leaves, stunted and spindly growth and reduced tillering. If deficiency persists to maturity, the number of grains/head is reduced. Deficiency appears first on older leaves; later on entire crop eventually will appear uniformly yellow.
Nitrogen fixation	The conversion of elemental nitrogen to organic combinations, or to forms readily utilizable in biological processes, by nitrogen-fixing micro-organisms. When brought about by bacteria in the root nodules of leguminous plants, it is referred to as symbiotic; if by free-living micro-organisms acting independently, it is referred to as non-symbiotic or free fixation.
Nitrogen-fixing plant	A plant which can assimilate and fix the free nitrogen of the atmosphere by the aid of bacteria living in the root nodules. Legumes with the associated rhizobium bacteria in the root nodules are the most important nitrogen fixing plants.
Nitrogen-free extract (NFE)	It consists mainly of carbohydrates like sugars and starches along with small amounts of pentoses and other less important carbohydrates.
Nitrophosphates	Products obtained by treatment of phosphate rock with nitric acid alone or in admixture with sulphuric or phosphoric acid, without subsequent treatment with ammonia.
Node	The part of a stem where the leaf or bud emerges.
Nodule	A structure developed on the roots of most legumes and a few other plants in response to the stimulus of root-nodule bacteria, usually genus *Rhizobium*.

Term	Terminology
Nomadism	In this system, the animal owners do not have a permanent place or residence. They do not practice regular cultivation and their families move with their herds.
Non-acid forming fertilizer	One that is not capable of increasing the residual acidity of the soil.
Non-discrete continuous variate	It is a variate which can take any value over a range of feasible values and these values are obtained by measurement.
Non-polar compounds	They are relatively uncharged molecules held together by van der Wall's forces and usually exhibit low water solubility and high oil solubility.
Non-saline-alkali soil	A soil that contains sufficient exchangeable sodium to interfere with the growth of most crop plants and does not contain appreciable quantities of soluble salts. The exchangeable sodium percentage is greater than 15 and the electrical conductivity is less than 4 millimhos per cm (at 25°C). The pH of the saturated soil paste is between 8.5-10.0.
Non-selective	A characteristic that enables a herbicide to become toxic and kill all plants treated.
Non-selective formulation	An herbicide that is generally toxic to all plants. Some selective herbicides may become non selective if used at very high rates.
Non-selective herbicide	A material that tends to kill plants with which it comes in contact.
Non-symbiotic nitrogen fixation	Fixation of molecular nitrogen by free-living micro organisms like *Clostridium* and *Azotobactor.*
Normal distribution	Symmetrical arrangement of replicate values that deviate randomly on either side of a mean value.
Normal ploughing (plough-sole depth)	Ploughing up to a depth of 15 cm.
Normal solution (N)	When equivalent weight of a compound is dissolved in litre of water the solution is one normal solution.
Normality of the solution	It is defined as the factor with which the equivalent weight must be multiplied in order to express the concentration of the solution.

Term	Terminology
No-tillage planting	Direct planting in essentially unprepared seed-beds.
Notorious herbicides	It refers to the herbicides which affect the crops in rotation even after two seasons of its application at normal rates.
Noxious weed	A weed plant, especially undesirable, troublesome— and difficult to control, *e.g., Cynodon dactylon, Cyperus rotundus, Sorghum halepense.*
	A weed arbitrarily defined by law as being especially undesirable, troublesome and difficult to control.
N-serve	A trade name for 2-chloro-6-(trichloromethyl) pyridine which is used as nitrification inhibitor. The rate is 1-2% N-serve by weight of fertilizer (ammonical) nitrogen.
Nuclear reactor	A device for supporting self-sustained nuclear reactions under controlled conditions.
Nucleic acid	A long molecule formed by lining many nucleotide molecules. The 2 kinds occurring in cells are DNA and RNA.
Nucleon	Any particle found as a constituent of the nucleus. Nucleoprotein A substance formed by the combination of protein and nucleic acid.
Nucleotide	A molecule formed by the union of phosphoric acid, a carbohydrate (pentose-ribose or deoxyribose), and a base derived from purine or pyrimidine (adenine, guanine, cytosine, thymine or uracil).
Nucleus seed	The original seed produced for the first time by the plant breeder. It is 100% pure in all genetical and physical quantities and used as a parent for the multiplication of breeder's stock seed.
Nuclide	Any one of the more than 1,000 species of atoms characterized by the number of protons and neutrons in the nucleus.
Null hypothesis	Hypothesis of no differences; that is there is no discrepancy between observation and expectation based on some set of postulates.

Term	Terminology
Nurse crop	A companion crop which nourishes the main crop by way of nitrogen fixation and/or adding the organic matter into the soil, *e.g.,* cowpea intercropped with cereals or new plantations of fruit trees.
	Such crop helps in the nourishment of other crops by providing shade and acting as climbing sticks.
Nursery	Series of plant progenies, introductions etc. grown chiefly for individual plant testing and examination in a plant breeding programme. The nursery bed where seedlings are raised for transplanting to the field.
Nut	An independent dry, 1 seeded fruit having a hard pericarp.
Nutrient cycling	The pathway of nutrient substances from their occunance in the physical environment to their incorporation in the living organisms and their return to the physical environment through the metabolic activity, death, and decay of organisms.
Nutrient-deficiency symptoms	When any essential plant nutrient is seriously lacking in the soil and plants growing on it show certain colour development in leaves and certain changes in their growth. These symptoms are characteristic to each of the nutrient. The nature of the symptom varies from crop to crop and with degree of deficiency.
Nutrition	The sum of the processes by which an animal or plant absorbs or takes in and utilizes food substances.
Nutritive value index (NVI)	Intake of feed DM × digestibility of DM × efficiency of utilization of digested nutrients.
Nutritive values	Nutritive value of forage is characterized by its chemical composition, digestibility and nature of the digested products. Chemical composition is a factor associated with only the plant and its environment, but digestibility, the nature of digested products and their efficiency of utilization are associated with both the plants and animals.
Oasis effect	It is the exchange of heat between a growing crop and Hot air whereby air over the crop is cooled.

Term	Terminology
Obligate aerobes	Microflora that grow only in the presence of molecular oxygen.
Obligate anaerobes	Microflora that grow only in the absence of molecular oxygen.
Obligate Parasite	A weed never found in the wild stage, but grows only in association with another plant, as a parasite.
Official seed sample	The samples which are drawn by a seed control or seed law-enforcement officer. These are submitted to the laboratory to determine if a seed lot being offered for sale, meets the requirements of the seed act.
Off-type	Plants or seeds deviating significantly from the characteristics of a variety as described by the breeder in any observable respect.
Oil (plant)	A slippery viscous liquid or fat like substances produced and stored in the seeds of plant species such, as sunflower, groundnut, soybean etc.
Oil cakes	When oil is extracted from oilseeds, the remaining solid portion is the oilcake. Oilcakes are of 2 types, edible oilcakes which can be safely fed to livestock and non-edible oilcakes which are not fit for feeding to livestock. Oilcakes are added to the soil as concentrated organic manures. They supply organic matter and all the major plant nutrients, mostly nitrogen.
Olericulture	A branch of horticulture which deals with cultivation of vegetables.
On-centre research	On-centre research aims to examine components of research that have important implication on technology development and is primarily multidisciplinary in nature.
On-farm research and development	Agronomic and socioeconomic studies conducted on the farms with farmer's active participation. The goals are to develop improved cropping system technologies and to devise ways to combine these technologies with farmer's knowledge and skill to efficiently utilize the available farm resources.

Term	Terminology
Ontogenic	Capable of producing or inducing tumors in animals, either benign or malignant.
Open pan evaporation (Eo)	It is the value of evaporation from an open pan (standard USWB), class A pan evaporimeter. Water needs of crops can be well predicted from the values of open pan evaporation in a given area and in a specified period.
Open-formula mixed fertilizers	Fertilizer mixtures of which the composition and the complete make-up are disclosed by manufacturers. Thus the fertilizer grade, the straight fertilizers and the filler used are disclosed on each bag. Cultivators and extension workers are in a better position to know the quantity and quality of nutrients contained in such a fertilizer mixture.
Operational Research Project (ORP)	It is an instrument of transfer of new technology. It is an integrated approach launched by the ICAR in co-operation with local agencies, voluntary organizations, agricultural universities, state agricultural departments to test, adopt and modify, if necessary, (i) research findings and make them suitable for large-scale adoption by farmers, (ii) to find out the constraints that impede acceptance and release of such findings, and (Hi) to find out the profitability of such findings.
Opportunity cropping	The practice of placing primary emphasis on the use of stored soil moisture while determining whether or not to establish a crop.
Optimum economic dose	If there is a quadratic response between input (X) and output (Y): $Y = a + bx + cx^2$, the optimum economic dose is given by: X opt $= q/p\text{-}b/2c$ Where q, cost of unit, x; p, price of unit y; b and c, constants.
Optimum tillage	A system of tillage contributing to maximum net return for crop tinder given field conditions.
Oral toxicity	Toxicity of a compound when ingested through the mouth.
Ordinate (Y-axis)	Vertical axis of a graph.
Organic	Organic materials those are not soluble in water.
Organic farming	It is a agricultural production system which avoids

Term	Terminology
(biological husbandry)	or largely excludes the use of synthetically compounded fertilizers, pesticides, growth regulators and livestock feed additives. To the maximum extent feasible organic farming systems rely upon crop rotations, crop residues, animal manures, legumes, green manures, mineral bearing rocks and aspects of biological pest control to maintain soil productivity and tilth to supply plant nutriments and to control insects, weeds and other pests.
Organic manures	Carbonaceous materials mainly of vegetable and/or animal origin added to the soil specifically for the nutrition of plants.
Organic matter	It includes plant and animal residues of various stages of decomposition in soil, cells and tissues of soil organisms, substances synthesized by soil microbial activity. Plant leaves, stem, straw, husk, roots, stubbles or any remnants of crops or plant species and organic manures including farmyard manure - are the cheap source of organic matter.
Organic nitrogen (soil)	Nitrogen in the form of organic compounds, *e.g.,* amino acids amino sugar, purine and pyrimidine bases.
Organic nitrogen fertilizer	Fertilizer containing nitrogen associated with carbon in organic combination.
Organic phosphorus	Phosphorus present as a constituent of an organic compound or a group of organic compounds, *e.g.,* glycerophosphoric acid, inosital phosphoric acid, cytidylic acid.
Organic soil	Soils containing organic matter in sufficient quantities to dominate the soil characteristics. Frequently. all soils containing 20% or more organic matter by weight are arbitrarily designated as organic soils.
Organoleptia	The term is used to describe the total sensory impressions associated with the spice, i.e., the aroma, flavour, pungency and bitterness sensations etc., imparted to the nose, mouth and throat.

Term	Terminology
Oriented tillage	Tillage operations which are oriented, in specific paths or directions with respect to the sun, prevailing winds, previous tillage actions or field base lines.
Orthogonal polynomials	When the effects represented by different polynomials of different degrees are independent, they are known as orthogonal polynomials.
Osmosis	The diffusion of water across a differentially permeable membrane from a region of greater water concentration to a region of lesser water concentration.
Osmotic effect	The force a plant must exert to extract water from the soil. The presence of salt in the soil water increases the force the plant must exert.
Osmotic potential	The amount of work that must be done per unit quantity of water in order to transport reversibly and isothermically an infinitesimal quantity of water from a pure water at a specified elevation and atmospheric pressure to a pool containing a solution identical in composition with soil water but in all other respects identical to reference pool.
Osmotic pressure	The pressure developed due to unequal concentrations of salts separated by a semipermeable membrane (or plant cell wall). Water will move from the side of lower salt concentration or higher free energy through the membrane to the side of high salt concentration or low energy. This water exerts additional pressure termed as osmotic pressure (numerically equal and opposite to solute potential).
Out crop	The actual edge of the inclined stratum at the surface ground.
Out cross	The mating of a hybrid with a third parent, also an off type plant resulting from pollen of a different sort contaminating a seed field.
Outer crop or guard crop or border crop	The crops which are grown around the field boundaries in narrow strips with twin objectives of protecting the main crop from stray cattle and producing livestock feed and/or seed are called outer or guard crops, *e.g., Sesbania* or *Leucaena* on boundaries of field/plantation crops and castor around spring-planted sugarcane.

Term	Terminology
Outlet channel	A channel which carries water out or away from a reservoir, lake or other water body.
Output parity index	It measures the parity of energy output to energy input. It combines both the economics and energetic approaches. Operationally it is the ratio of price per unit of energy output to the price per unit of energy input.
Ovary	The enlarged basal part of a pistil which contains an ovule or ovules (potential seeds).
Oven-dry soil	Soil is considered to be oven dry when it has reached equilibrium with the vapour pressure from an oven at 105°C. The oven-dry condition corresponds to a relative humidity of approximately 0% of pF near 7.
Over irrigation	The application of more water than is necessary for the needs of vegetation, resulting in loss of water through seepage and leaching.
Over stocking (animal)	Stocking animals beyond the safe grazing capacity of a pasture.
Over yielding	Production of component crops in an intercrop which is higher than the sum, of appropriate monoculture crops. This is indicated by an LER greater than unity.
Oversee ding	It refer to seeding of a legume or grass into an existing pasture in order to improve the productivity and quality of herbage.
Ovule	A structure which develops into a seed after the contained egg is fertilized.
Oxidation	It is an energy-yielding process usually resulting from loss of hydrogen or addition of oxygen to a compound. The electron acceptor is called an oxidant.
Oxisols	Soils with residual accumulations of inactive days, free oxides, kaolin and quartz. They are mostly found in tropical climates.
Oxylophytes	Plants tolerant to high acidic soil conditions.
Paddock	Small fenced field used for grazing purposes.
Paddy sheller or dehusker	An equipment to remove husk from paddy grain.

Term	Terminology
Paddy weeder	Equipment for inter culture used in paddy cultivation. It is used for uprooting weeds and burying them in puddled soil between rows of standing crop. It improves the aeration of the soil.
Padzol soil	A soil having an organic mat and a very thin organic mineral layer above a grey leached layer, formed under a forest in a cool humid climate.
Pair production (nuclear)	The transformation of high-energy proton to an electron and a positron.
Paira crop (uthera)	It is a crop sown broadcast in the standing crop of lowland rice before its harvest where the residual moisture is used for the establishment of uthera crop, *e.g.,* Lathyrus, gram, lentil, etc. in standing crop of rice.
Paira/Utera crops	Growing of such crops sown a few days or weeks before harvesting of standing crops is called paira/Utera cropping and the sown crop is called paira/utera crop.
Paired row cropping	Each third row is removed or growing of crops in paired row is called paired row cropping. It is suitable for dry land and objective is to conserve soil moisture.
Paired row planting	In this technique 2 adjacent rows of the base crops are paired, reducing the inter-row space in the pair narrow enough to create some interspaces between pairs of base crops rows but wide enough to minimise undue competition among plants.
Palatability	Plant characteristics eliciting a choice between two or more forages or parts of the same forage, conditioned by the animal and environmental factors that stimulate a selective intake response
Palea	The upper of the 2 bracts enclosing each floret in the grass spikelet.
Pan	Pan is a layer formed by accumulation of materials such as salts, clay, etc., and which impedes free drainage. In a clay pan, there is accumulation of clay washed down from upper layers. Dissolved salts like silica, calcium carbonate, etc. on precipitation, form a hard pan.

Term	Terminology
Pan coefficient (kp)	Ratio between reference evapotranspiration ETo and water loss by evaporation from an open water surface of a pan; k = ETo/E pan; fraction.
Pan evaporation	Is the value of evaporation from an open pan evaporimeter, usually from standard USWB, class A pan evaporimeter during a day or specified period at a place.
Panicle	An inflorescence having repeated branching, with each branch bearing a flower in a cereal crop or that portion of the plant that bears seeds.
Panicle-initiation stage	The stage begins when the primordia of panicle has differentiated and becomes visible.
Parameter	A characteristic of a statistical population, such as its average, its standard deviation or variance and are often cannot be measured directly but are estimated from samples of the population.
Parasite	A plant or animal which lives in or on or with some other living organism which is called its host.
Parasitism	A relationship in which a heterotrophic organism, obtains its food from the living tissues of another organism with no recompense to the host, *e.g.,* orobanche on tobacco.
Parboiling (rice)	Quite often the paddy is subjected to parboiling treatment before milling and it involves soaking the paddy in water, steaming it in hot water and drying it in the sun. Parboiling makes husk-removal easier toughens the grain which results in reduced breakage during milling and polishing and improves the keeping quality of rice.
Parent material	The horizon of weathered rock or partly weathered soil material from which the soil is formed. Horizon 'C' of the soil profile.
Parshall flume	A calibrated device for measuring the flow of water in open conduits. It consists of a contracting length, a throat and an expanding length.
Part per million (ppm)	Designates the quantity of a substance contained in a million parts of a mixture or solution in carrier, such as air or water.

Term	Terminology
Parthenocarpy	The development of fruit without fertilization or the phenomenon of seedless fruit production.
Partial correlation	Partial correlation coefficients measure the strength of the linear relationship between one dependent variable and one of several independent variables after controlling the effects of the other independent variables by holding them statistically constant.
Particle density (soil)	Particle density is defined as the ratio of mass of solids (oven-dried soil) to the volume of the solids alone and is expressed as g/cm^3.
Passive absorption of water	The absorption of water by the roots as a result of forces originating in the leaves, such as transpiration pull, and does not require any expenditure of metabolic energy.
Pasturage	Vegetation on which animal grazes, including grasses or grass-like plants, legumes herbs and shrubs.
Pasture	The term pasture is applied to a grazed plant community usually composed of several species often of diverse botanical type.
Pasture land	An area of land covered with grass or other herbaceous forage plants used for grazing animals.
Pasture renovation	Improvement of a pasture by tillage, seeding, fertilization and sometimes by liming also.
Pasture succession	A series of crops for grazing in succession.
Peak or maximum supply (V max)	Average daily supply requirement during the peak water-use period for a given crop or cropping pattern and climate m^3/day.
Peak supply period (water)	Water-use period for a given crop or cropping pattern during the month of period thereof of highest water requirements; mm/day.
Pearling	This is another term used for polishing or whitening food grains.
Peasant farming	The ownership and operation ship both are individual. The farmer, i.e., owner of land in this 'case is the worker and manager both. In peasant farming high premium is placed on management and the use of family labour.

Term	Terminology
Peat (soil)	Unconsolidated soil material consisting largely of undecomposed or only slightly decomposed organic matter accumulated under conditions of excessive moisture.
Ped	A unit of soil structure such as an aggregate, crumb, prism, block, or granule, formed by natural processes (in contrast with a cold, which is formed artificially).
Pedicel	The stalk of 1 flower in a cluster.
Pedigree breeding	System of breeding in which individual plants are selected in the segregating generations from a cross on the basis of their desirability judged individually and on the basis of pedigree record.
Pedology	The science dealing with the laws of origin, formation and geographic distribution of the soil as a body in the nature.
Peduncle	The main stalk of an inflorescence or the stalk of a solitary flower.
Peeling	The removal of the outer layer of a fruit or vegetable is generally referred as peeling or skinning.
Pelleted formulation	A dry formulation consisting of discrete particles usually large than 10 cubic millimeters and designated to be applied without a liquid carrier.
Pelleted seed	Layer of particular matter around the seed, pelleting is usually made from finely pulverized limestone or mineral phosphate.
Pellicular zone	The maximum depth from the natural surface up to which the evaporation can have its effects.
Penning	It is a practice of confining cattle by erecting a small enclosure in the field at nights to harvest the solid and liquid excreta of the cattle.
Per cent germination	The proportion of the number of seeds which have produced seedlings classified as normal under the conditions and period specified.
Percentage of phosphorus derived from fertilizer (P diff %)	It is calculated by using the formula: P diff % = Specific activity of plant ÷ sample Specific activity of fertilizer standard × 100.

Term	Terminology
Perched water-table	The upper limit or surface of a small body of water above the main water-table. The water is retained in its elevated position by an impervious stratum and may form a limited source of water supply.
Percolation	It is a downward movement of water through the interstices of rock or soil under gravity or hydrodynamic pressure or both under saturated or nearly saturated condition.
Perennial	Plant that dies back seasonally and produces new growth again from a perennial part.
Pericarp	The outer protective wall that covers the kernels of cereal grains. It consists of several layers, the principal ones being endocarp, mesocarp and epicarp.
Periderm	A part of the bark (of woody plants) that contains tightly packed supersized cells (with no intercellular spaces) which are composed of fatty acids, lignin, cellulose and terpenes; it is composed of phellogen, the phellem (cork) and the phelloderm.
Permanent cropland area	Land cultivated with crops that need not be replanted after each harvest, such as rubber, coffee, cocoa, including land under shrubs, fruit trees, nut trees and vines but excluding land under trees grown for food or timber.
Permanent cropping system	An agricultural system in which crop plants are grown continually on the same plots of land, either several crops in rotation or a single crop for several years; disturbance of microclimate and of water relations is permanent; nutrients and organic matter are replaced by inputs from outside.
Permanent pasture	Pasture of perennial or self-seeding annual plants maintained for several years for grazing.
Permanent wilting point	The moisture content in percentage of a soil at which nearly all plants wilts and does not recover in a humid dark chamber unless water is added from an outside source. This is the lower limit of available moisture range for plant growth. Below the wilting point extraction of moisture continues for some time but growth ceases completely. The force with which moisture is held by the soil at this point corresponds to 15 atmospheres.

Term	Terminology
Permanently absorbed water	Water which passes downward beyond the root zone, finally reaching the zone of saturation where it becomes an increment to groundwater.
Permeability	The property of a porous medium to transmit fluids. It is a broad term and can be further specified as hydraulic conductivity and intrinsic permeability. 'K' in Darcy's law represents the rate of flow of water in a porous medium under a unit hydraulic gradient.
Permeability coefficient	The rate of flow of fluid through a unit crosssection of a porous mass under a unit hydraulic gradient at 60°F temperature.
Permeability rate	The rate at which water moves through a soil under a standard pressure gradient, usually defined as amount of water traversing per sq cm of soil surface under a pressure gradient of one dyne per cm.
Permeameter	A device for measuring the permeability of soils or other materials.
Persistence	Ability to withstand extremes of climatic, edaphic, and biotic stresses in nature.
	Persistence is the perennation of the original plants, and maintenance of the species by vegetative spread or by regeneration from self-sown seed.
Persistent herbicide	A herbicide which, when applied at the recommended rate will harm susceptible crop(s) planted in normal rotation after harvesting the· treated crops or which interferes with regrowth of native vegetation in non-crop sites.
Pesticides	They comprise a large number of various chemical substances used to control arthropod pests and infectious diseases, improve the fertility of soils, and prevent the growth of weeds. They include insecticides, herbicides, nematicides and fungicides.
Petrophyte	A plant able to grow on rocks.
pF	It is the logarithm of height in cm of a water column that represents total stress with which water is held by a soil.

Term	Terminology
pH	Refers to the potential of hydrogen, defined as the negative logarithm of the hydrogen ion concentration of a solution or pH = -log (H$^+$), where (H$^+$) = hydrogen ion activity.
pH hydrolytic	The arithmetical difference between the pH value of a soil as measured on the soil paste and the value obtained on a 1:10 suspension.
pH isohydric	The pH value of a soil identical with that of a buffer solution which remains unchanged when mixed with the soil.
pH soil	The negative logarithm of the hydrogen-ion activity of a soil. The degree of acidity (or alkalinity) of a soil as determined by means of a glass quinhydrone or other suitable electrode or indicator at a specified moisture content or soil to water ratio and expressed in terms of the pH scale, from 0 to 14.
pH Value	An expression of the degree of acidity or alkalinity, related to the number of hydrogen (H) ions in a water solution. pH values below 7.0 indicate acidity with its intensity increasing as the number decrease, and conversely, pH values above 7.0 indicrease alkalinity with its intensity increasing as the numbers increase
Phased planting (staggered planting)	It is planting or sowing of a crop spread over an optimum period of planting either to minimize risks or to use labour or machinery more effectively or to minimize competition (in intercropping) or to prolong the period of supply to the market or to the factories.
Phenogram	Diagram representing the variation of a given phenological element in terms of meteorological factors.
Phenology	The study of the timing of recurring biological events, the causes of their timing with regard to biotic and abiotic factors and the interrelation among the phases of the same or different species.

Term	Terminology
Phenophases (growth stages)	The stages of development in the life-cycle of a plant in relation to the climatic conditions. The important phenophases in cereal crop plants are emergence, tillering or branching, heading, flowering, anthesis etc.
Phenotype	Physical expression or outward appearance of a genotype which is subject to change with the environment.
	The visual characteristic of an organism or plant as determined by the interaction of its genetic constitution and environment.
Phloem	The principal food-conducting living tissue of the vascular plant, basically composed of sieve tube, companion cells, fibers and sclereids; transports metabolic compounds from the site of synthesis or storage to the site of utilization.
Phosphate mineralization	Solubilization of insoluble organic phosphatic compounds (*e.g.,* phytin, lecithin, etc.) by micro-organisms by virtue of their organic acid production.
Phosphate potential	The amount of work that must be conducted to move reversibly and isothermally an infinitesimally small amount of a phosphate ion from a pool of phosphate at a specified location at atmospheric pressure to the point under consideration. It is an index for availability of soil phosphorus.
Phosphate rock	A natural rock containing one or more calcium phosphate minerals of sufficient purity and quantity to permit its use directly or after concentration in the manufacture of commercial phosphorus fertilizers.
Phosphate solubilization	Solubilization of insoluble inorganic phosphates (tricalcium phosphate, rock phosphate etc.) by micro-organisms by virtue of their organic acid production.
Phosphates	Various salts of phosphoric acid applied to the soil as phosphatic fertilizers to supply phosphorus to crop.

Term	Terminology
Phosphatides	These are a group of substances of fatty nature which are essential components of plant cells and are usually removed from edible oils. They contain phosphorus and nitrogen and on hydrolysis give fatty acids, phosphoric acid and basic substances such as choline.
Phospholipids	Substituted lipids containing phosphoric acid and nitrogen, for example lecithin, cephalin etc.
Phosphor (nuclear)	A material such as zinc sulphide which gives of visible light when struck by nuclear radiation.
Phosphorus deficiency	Deficient plants usually have dark green, more erect leaves than normal. In some crops leaves show orange or purplish discolouration. Phosphorus deficiency can occur in strongly acid, acid sulphate, peat and alkali soils, and is more available in flooded soils than in dryland soils.
Phosphorylation	The process by which high-energy organic phosphate compounds are produced, using respiratory or sunlight energy.
Photo synthetically active radiation (PAR)	Ecologist and others frequently give their measurements of irradiance in energy units for wave-lengths from 400 to 700 nm, the wavelengths most active in photosynthesis. Appropriate SI unit for PAR are watts/m^2 (W/m^2).
Photocathode	A device that liberate electrons when struck by photons of light.
Photochemical reaction	A chemical reaction which proceeds at the expense of absorbed light, for example photosynthesis.
Photodecomposition	Decomposition or detoxification of herbicides by exposure to light in the soil is called as photodecomposition.
Photo-inductive cycle	The photoperiodic cycle that induces flower initiation or brings about other responses.
Photomorphogenesis	A reaction in which the plant shows changes in growth and development caused by visible light, independent of photosynthesis.
Photon	A quantum of electromagnetic radiation.

Term	Terminology
Photo-oxidation	When the light intensity exceeds a limit, leaves consume oxygen and use it in the oxidation of certain cell constituents. It is a chlorophyll-catalysed process and more chlorophyll molecules become excited than possibly is utilized causing harmful side effect. It results in bleaching of chlorophyll and inactivation of important enzymes.
Photoperiod	The relative length of time that a plant is exposed to light especially as it affects the life cycle and physiological process of the plant.
Photoperiodism	The effect of day-length on flowering, dormancy, vegetative development, leaf fall and other aspects of plant behaviour.
Photophosphorylation	Production of high-energy phosphate (ATP) from inorganic phosphorus and ADP through the utilization of light energy in photosynthesis.
	The process of ATP formation, in plants, in the presence of light.
Photorespiration	Light-dependent oxygen uptake and carbon dioxide production (distinct from mitochondrial respiration), and is primarily more in C_3 plants and is very rare in C_4 plants.
Photo-spectrometer (emission spectrometer)	An instrument that disperses the light from a source and produces a spectrum indicating relative intensity as a function of wavelength.
Photosynthesis	Conversion of light energy into chemical energy by plants with the help of chlorophyll, carbon dioxide and water.
	The process that converts atmospheric carbon dioxide and water to carbohydrates, using sunlight as the source of energy, in plants.
Photosynthetic efficiency	The percentage of total solar energy assimilated by plants. The energy fixed by plants expressed as percentage of the incoming solar energy.
Photosynthetic number	The number of grams of CO_2 absorbed per hour per gram of chlorophyll.
Photosynthetic quotient (PQ)	The ratio of volume of O_2 evolved to the volume of CO_2 absorbed in photosynthesis or $PQ = O/CO_2$.

Term	Terminology
Phototropism	Growth response of plants to variation in light intensities.
Phreatic rice lands (upland or lowland rainfed)	Whether they are naturally sloping or flat, mayor may not be bunded and free ground-water is present within the root zone of rice plant during the growing season. It differs from pluvial with regard to water table depths.
Phreatophyte	A plant that indicates the presence of subsurface water.
Physical determinants	The important attributes of climate, water, and land such as rainfall, topography, and hydrology that influence configuration and performance of cropping patterns.
Physiological adaptation	Plants may have physiological adaptations which result in resistance to parasites, greater ability to compete for nutrients, or ability to withstand desiccation.
Physiological Antagonism	The type of antagonism that occurs when two components of a mixture, acting at different sites, counteract each other by producing opposite effects on the same physiological process.
Physiological drought	Non-availability of water to the plant due to unfavourable physiological conditions, such as water logging, soil salinity, low temperature etc.
Physiological maturity	Physiological maturity refers to a developmental stage after which no further increase in dry matter occurs in the economic part.
Phytic acid	It is an inositol hexaphosphoric acid. The acid is important in nutrition as it precipitates calcium and prevents its absorption by the body.
Phytobland oil (crop oil)	Non-phytotoxic oil used to aid herbicide penetration in plant.
Phytochrome	A photo-morphogenic protein pigment in plants involved in photoperiodic response of plants. It is present in 2 inter-convertible forms, the red light-absorbing form (pr) and other far-red light absorbing form (pfr).

Term	Terminology
Phytoclimatology	Study of microclimate in relation to plants.
Phytomass	Organic matter produced by green plants grown on land or in water through the process of photosynthesis by direct utilization of sun's energy.
Phytotoxic	Injurious to plant life or life processes.
Piche atmometer	A porous paper wick atmometer for measuring evaporative rate of water.
Pictogram	A form of graphical presentation of numerical data by drawing pictures of the particular object referring to the data.
Pie diagram	It consists of a circle cut into sectors, the areas of the sectors being proportional to the magnitudes of the component parts.
Piezometer	Instrument used for measuring depth of water table.
Pigments	Substances appearing coloured due to differential absorption of light such as chlorophyll, carotenoids, etc.
Pilot production programme	A small-scale (100 to 500 ha) production to determine the support needed in the large-scale diffusion of recommended technologies as well as to clearly specify the tasks and interrelationships of different institutions involved in supporting the farmers. It also allows final evaluations of the recommended cropping systems technology, the cost of its extension to the farmers, and the expected benefit.
Pitcher farming	A practice in dry farming where crop is irrigated through small holes made in the bottom of earthen pitcher. The practice is generally used for wider spaced plants.
Placement (fertilizer)	Inserting, drilling or placing the fertilizer below the soil surface by means of any tool, implement or equipment at desired depth to supply plant nutrients to crop before sowing or in the standing crop.
Plagiotropic	The term used primarily for roots, stems or branches to describe growth at an oblique or horizontal angle.
Planking	One of the tillage operations carried out before sowing the crop for smoothening, microlevelling and to cover the seeds after sowing.

Term	Terminology
Planophile	Prostrate-leaved canopy which intercepts more light per unit of LAI.
Plant breeding	The applied branch of botany dealing with the crop improvement and production of new improved crop varieties far better than original in all aspects.
Plant ecology	The branch of ecology which deals with the study of plants in relation with the environment.
Plant food ratio	The ratio of the number of fertilizer units in a given mass of fertilizer expressed in the order NPK.
Plant geometry	The pattern of distribution of plants over the ground or the shape of the area available to the individual plant.
Plant growth regulator	A substance used for controlling or modifying plant growth processes without severe phytotoxicity
Plant growth regulators	Organic compounds occurring naturally in plants as well as synthetic other than nutrients, which in small amounts promote inhibit or modify any physiological process in plants.
Plant interference	The response of an individual plant or group of plants to its total environment as modified by the presence and or growth of other individual or species of plants. It denotes total mechanism whereby a plant or a group of plants suppresses or otherwise modified the growth of neighbouring plants.
Plant introduction	The process of introducing the plants from their growing locality to a new locality, having a different climate.
Plant nutrient	A chemical element essential for plant growth. Plant pathogen: Any organism capable of causing disease in a particular plant host or range of host plants.
Plant physiology	The study of plant processes, functions, and the responses of living plants in relation to environment.
Plant stand or plant population or plant density	The number of plants per unit area in a crop field.
Plant tissue test	This refers to the rapid chemical testing of plant tissue for colorimetric determination of the levels of nitrate, phosphorus and potassium in the sap of fresh plant tissue. Sometimes, tests are also carried out for magnesium, calcium, manganese and zinc.

Term	Terminology
Planter	A machine which is used for precision drilling, hill dropping or check row planting.
Planting ratio (seed production)	It is recommended ratio in which male and female parental lines are planted to make a crossing block in hybrid seed production.
Plastic mulch	Thin polyethylene film which maybe transparent or black that is used as mulch for field crops and helps in moisture retention, increasing soil temperature and weed control.
Plot (experimental)	An experimental unit to which a treatment is applied (it may be single plant, small area of soil containing several plants, a strip through a field or a whole field).
Plough sole	A compacted layer at the bottom of the zone of ploughing.
Plough sole placement	A method of fertilizer application in which fertilizer is placed in a continuous band in the bottom of the furrow, in the process of ploughing each band is covered as the next furrow is turned.
Ploughing	Operations carried out with the help of tractor drawn or bullock-drawn implements known as plough, before the crops are sown.
Plumule	The major young bud of the embryo within a seed from which aerial portions of the plant develop. It usually occurs at the tip of the epicotyls, the part of the embryonic plant axis above the cotyledons.
Pluvial rice lands (rainfed uplands)	Lands which are well drained without free ground water within the rooting zone of the rice plant and are generally unbunded.
Podzol soil	A soil having an organic mat and a very thin organic mineral layer above a grey leached layer, formed under a forest in a cool humid climate.
Podzolization (silication)	The chemical migration of aluminium and iron and/ or organic matter, resulting in the concentration of silica (*i.e.* silication) in the layer eluviated.
Poisson distribution	Discrete probability mass function involving events which take place rarely' or relatively infrequently when only a small subinterval of time is being considered.

Term	Terminology
Polar compounds	Characterized by electrical charges.
Polarimeter	It is equipment used for measuring Pol reading for computing the reducing and non-reducing sugars in sugarcane juice.
Polarity	The electrical phenomenon of a molecule or anion., on the basis of which the chemicals can be divided into polar and nonpolar compounds.
Polishing	The removal of bran layer from brown rice by mechanical operation. It is also known as whitening or pearling.
Pollination	The transfer of pollen from anther to a stigma.
Pollution	Contamination of natural environment by the addition to air or water of substances potentially toxic or otherwise harmful to man and animals for example, SO_2 CO, radioactive fallout, insecticides etc.
Polynutrient fertilizer	A fertilizer containing more than one major plant nutrient. It is synonymous for multiple nutrient material and complex fertilizers.
Polyphosphate	It is the most concentrated form of phosphatic fertilizer and is very suitable for making liquid fertilizers. It is obtained by neutralizing super phosphoric acid (about 75% P_2O_5) with ammonia; the product being called ammonium polyphosphate (containing 15% N and 58% P_2O_5).
Population	A hypothetical and infinitely large series of potential observations among which observations actually made constitute a sample. It is the collection of data that we wish to investigate.
Pores (macro)	When diameter is 100 microns or more, main functions are to maintain air circulation and drainage of soil.
Pores (meso)	Pore size diameter ranges between 30 and 100 microns and helps in capillary water conduction or movement.
Pores (micro)	Diameter of pores ranges from 3 to 30 microns; these play important role in water retention (slow capillary flow).

Term	Terminology
Porosity (pore-space)	It is the percentage of soil volume not occupied by the soil particles.
Post-emergence	Any treatment after emergence of the specified weed or crop plants.
Post-emergence treatment	Treatment is made after emergence of specified weed or crop.
Post-harvest tillage	The cultivation of the land immediately after harvesting at physiological maturity to aid in the preparation of the land for the next crop.
Potash (potassium oxide, K_2O)	The potassium content of a potassic fertilizer.
Potassium absorption ratio (PAR)	A ratio for soil extracts and irrigation waters used to express the relative activity of potassium ions in exchange reactions with soil.
Potassium chloride	Commercial potassium chloride is a potash salt that contains not less than 48% potash, chiefly as chlorides. Normally, potassium chloride or muriate of potash sold as fertilizer in India contains 60% K_2O.
Potassium deficiency	Symptoms of mild deficiency are dark green leaves, low tillering and stunting. Severe deficiency results in yellowish-orange to yellowish-brown discoloration, starting at the tip of older leaf blades and gradually moving towards base. Necrotic spots may appear on the leaf blade. Grain size and weight may be reduced.
Potassium fixation	The process of converting exchangeable or water-soluble potassium to a form not easily exchanged from the adsorption complex with a cation of a neutral salt solution.
Potassium magnesium sulphate	A double salt of potassium and magnesium containing about 21% K_2O.
Potassium nitrate (KNO_3)	A nitrogen-potassium fertilizer containing 13 per cent N and 44 per cent K_2O.
Potassium sulphate (K_2SO_4)	A potassium fertilizer containing 48% K_2O and also supplying 17-20% of sulphur.

Term	Terminology
Potential acidity	The amount of exchangeable hydrogen ions in a soil that can be rendered free or active in the soil solution by cationic exchange.
Potential crop yield	Yield of a crop obtained at research stations under most ideal environment.
Potential digestibility	It is defined as the maximum digestibility attainable when conditions and durations of fermentation are not limiting factors. The potential digestibility of some plant fractions may be determined with animals since the fractions are completely or almost completely digested.
Potential evaporation	It represents evaporation from a large body of free water surface. It is assumed that there is no effect of adjective energy. It is primarily a function of evaporative demand of climate.
Potential evapotranspiration	Maximum quantity of water capable of being evaporated in a given climate from a continuous stretch of vegetation covering the whole ground and well supplied with water. It thus, includes evaporation from the soil and transpiration from the vegetation of a specified region in a given time interval expressed in depth units.
Potential matric	It is the portion of the water potential that is attributable to the more or less solid colloidal matrix of the soil or plant system. It is the amount of work that a unit quantity of water in an equilibrium soil-water (or plant-water) system is capable of doing when it moves to another equilibrium system identical in all respects that there is no matric present.
Potential pressure	It is the portion of the water potential that results from an overall pressure, that is different from the reference pressure; it is the amount of work that a unit quantity of water in an equilibrium soil water (or plant-water) system is capable of doing when it moves to another equilibrium system identical in all respects except that it is at reference pressure.

Term	Terminology
Potential soil water	It is the amount of work that must be done per unit quantity of pure water in order to transport reversibly and isothermally an infinitesimal quantity of water from a pool of pure water at a specified elevation and atmospheric pressure to the soil water at the point under consideration.
Potential solute (osmotic)	It is the portion of the water potential that results from the combined effect of all solute species present in the soil or plant system. It is the amount of work that a unit quantity of water in an equilibrium soil-water (plant-water) system is capable of doing when it moves to another equilibrium system identical in all respects except that there are no solutes.
Potential total	It is the sum of all potentials acting on water in an equilibrium system. It is the amount of work that a unit quantity of water in an equilibrium soil-water (plant-water) system is capable of doing when it moves to a pool of pure free water at the same temperature located at a reference level and subject to atmospheric pressure.
Potential water	It is the difference between chemical potential of water in an equilibrium system and the chemical potential of water at the same temperature in the reference state. Thus it is the sum of all components of chemical potentials for isothermal condition, i.e., it is the sum of the pressure potential, the matric potential and solute potential. The water potential is the amount of work that a unit quantity of water in an equilibrium soil water (or plant water) system is capable of doing when it moves to a pool of water in the reference state at the same temperature.
Potential, gravitational	A potential attributable to gravitational force field and is dependent on the elevation or vertical location of the water. It is the amount of work that a unit quantity of water in an equilibrium soil-water (or plant-water) system at an arbitrary level is capable of doing when it moves to another equilibrium system identical in all respects except that it is at a reference level.

Term	Terminology
Poverty adjustment	The range of supply of a limiting element throughout which an increase in that element brings about an increase in plant growth. There is no increase in per cent of the element in the plant.
Power tiller	A walking type of tractor. This tractor is usually fitted with 2 wheels only. The direction of travel and its control for field operations is performed by the operator walking behind the tractor.
Prairie	The level or rolling area of treeless land covered with grass.
Precipitated bone phosphate	A by-product from the manufacture of glue from bones obtained by neutralizing the hydrochloric acid solution of processed bone with lime. The phosphorus is chiefly present as dicalcium phosphate.
Precipitation	Total amount of precipitation (rain, drizzle, snow, hail, fog, condensation, hoar frost and rime) expressed in depth of water which would cover a horizontal plane if there is no runoff, infiltration or evapotranspiration; mm/day.
Precision	It denotes relative or apparent nearness to the truth.
Precision drilling (seed)	Uniform placing of seeds in rows at a predetermined depth and seed rate.
Preemergence	Any treatment after a crop is planted but before emergence of the specified weeds or crop plants or both.
Pre-emergence treatment (herbicides)	Herbicides applied after seeding but before the crop or weed emergence.
Pregerminated seed	The procedure of germinating seed before sowing, usually until the radical just emerges, as practiced in direct-seeded rice.
Preparatory tillage	Various tillage operations carried out on the farm before the crops are sown, planted or transplanted to make a fine seedbed.
Preplant Application	Any treatment applied before planting (seeding) or transplanting a, crop, either as a foliar application to control the 'existing vegetation or as a soil application.

Term	Terminology
Preplant Incorporation	Any treatment applied and incorporated (blended) into the soil before seeding or transplanting a crop or weed; incorporation usually by tillage.
Preplanting treatment (herbicides)	Treatment applied after the soil has been prepared but before seeding or planting.
Preproduction evaluation	Higher level on-farm cropping system activities consisting of multilocation pattern testing and pilot production programme to delineate the final production programme area, verify technology performance, and determine institutional Support requirements and to help in structuring large-scale production programmes.
Press mud	It is one of the by-products of sugar industry, used as manure in the field; contains about 1.25% N, 2% P_2O_5 and 20–25% organic matter. Since it is very high in lime (up to 45%), its application is useful in acidic soils.
Pressure gradient	The rate of decrease of pressure in space at a given time.
Prey crop (smother crop)	The crop which is grown for the purpose of eliminating any undesirable plant through physical or its allelopathic effect.
Prill	Particle obtained by solidification of falling droplets of fertilizer.
Primary growth	Growth which is initiated by apical stem and root meristems.
Primary metabolites	Substances that are found essentially in all plants. These play reasonably well understood roles in the physiology of the plant, for example carbohydrates and proteins.
Primary plant foods (major nutrients)	These are nitrogen (N), phosphoric acid (H_3PO_4) and potash (K_2O).
Primary productivity	This is the productivity of pasture or grassland and is measured in terms of herbage accumulation per unit area per unit time. This is a direct measurement of pasture productivity.

Term	Terminology
Primary tillage	Tillage operations which constitute the initial major soil working operation. It is normally designed to reduce soil strength, cover plant materials and rearrange aggregates.
Priming	A term is used to designate the harvesting of ripe tobacco leaves from the stalk as they ripen beginning at the bottom and progressing upward. It also means priming water into a pump to prepare for its operation.
Primordial stage	Initiation or first formation of earhead in quiet primary reproductive stage in plant body. The primordia formation occurs after the maximum tillering stage in rice, wheat etc.
Principle of limiting factor (Liebig's law of minimum)	It states that when a process is conditioned as to its rapidity by a number of separate factors, the rate of the process is limited by the factor which is at the minimum.
Probability	The proportion of times in which an event occurs in an infinitely large and hypothetical series cases, each capable of producing the event.
Probability distribution	It is defined as a list of all possible outcomes to an experiment along with the associated probability of occurrence of each of the outcomes.
Probability value	A number between 0 and 1 that indicates how likely an event is to occur.
Processing	A part of marketing service which deals with the conversion of the produce into a more finished condition before it is offered for sale.
Production complex	A union of sites where relative performance of cropping patterns is substantially same. It may contain more than one environmental complex.
Productivity (plant)	The capacity of a plant to produce yield or biomass per unit time.
Progeny testing	A method of ascertaining the genotype of a plant by studying the appearance of the offspring from controlled crosses.
Project efficiency (Ep)	Ratio between water made directly available to the crop and that released at project headworks; Ep = Ea. Eb. Ec; fraction.

Term	Terminology
Pro-modernity index	It measures the extent to which energy input in the-farm production process comes from the goods developed through the use of modern technology or manufacturing process such as farm machinery, commercial fertilizers, insecticides etc.
Properties (chemical)	It is characterized as chemical composition of soil or medium, such as contents of organic carbon, phosphorus, potash, pH, electrical conductivity etc.
Properties (physical)	It is characterized as physical composition of soil or medium, such as sand, silt, clay, texture, structure etc.
Proportional counter (nuclear)	A radiation-detector designed to provide linear gas amplification of the primary ionization, thus providing an output pulse whose height is proportional to the energy of the incident radiation.
Proprietary cultivars	Cultivars breed, registered and produced by private companies.
Protein efficiency ratio (PER)	A term used for expressing the growth-promoting ability of a protein, i.e., gain in weight (g)/g protein consumed.
Protein synthesis	An elaborate process of protein formation in plant cells consisting of activation of amino acids; attachment of amino acids activated to s-RNA and formation of polypeptides on ribosome. Transcription and translation are the processes involved at molecular level in the passage of message from DNA.
Proteins	A class of foods composed of carbon, hydrogen, oxygen and nitrogen and often sulphur and phosphorus. These are made of repeating amino acid units linked together by peptide bonds.
Proton	A positively charged elementary particle of an atom having a mass number of 1 (the nucleus of a hydrogen atom of mass 1) and a of an atom mass of 1.67×10^{-24} g.
Protoplasm	The living matter in plant and animal cells, referred to as the physical basis of life.
Protoplast	The cell excluding of cell wall.

Term	Terminology
Prussic acid (HCN, Dhurin)	A highly toxic gaseous compound, found in certain crops like sorghum, cassava, etc. and in some poisonous weeds.
Psammophyte	A plant that grows on sand.
Psychrometer	Instrument used to measure the humidity of the atmosphere. It comprises 2 identical thermometers– the bulb of one is dry and the bulb of other is covered with thin film of distilled water.
Psychrometric constant	It is the ratio of specific heat of air to the latent heat of evaporation of water.
Psychroxerophyte	A drought-resistant plant growing in cold territories.
Pubescent	Presence of hair on stems or leaves; it may affect wetting of the foliage and retention of spray.
Puddle	A compact mass of earth, soil, clay or mixture of 2 or more of them which has been compacted through the addition of water and rolling and trampling and made more or less impervious.
Puddled soil	Dense massive soil artificially compacted when wet and having no regular structure. The condition commonly results from the tillage of a clayey soil when it is wet.
Puddler	An implement used for preparation of rice fields with standing water after initial ploughing. It breaks up clods and churns the soil. The main purpose of puddling is to reduce leaching of water.
Puddling	Making a soil impermeable by manipulating and compacting it in standing water, which reduces its apparent specific volume thus facilitates transplanting.
Pull	The total force required to pull an implement.
Pulse seed	The edible seed of legumes.
Pulverization	An operation of mechanically turning down of upper surface soil or materials by means of tools or tillage implements.
Pungency	The hot sensation produced on the tongue by certain constituents of some spices, *e.g.,* by the gingerols in ginger, by piperine in pepper and by the capsicinoids in *Capsicum* species.

Term	Terminology
Pure line selection	The isolation of desirable homozygous plants from the mixed population and multiplying without contamination to release as improved variety.
Pure seed	The seed true to its kind or variety.
Purity test	It refers to the determination of the percentage purity of the item, *viz.* pure seed obtained after discarding other crop seed, weed seed, inert matter, based on examining the quantity Specified under rules.
Put-and-take animals	Used in grazing experiment to graze excess forage beyond that needed for tester animals and to accumulate animal days.
Put-and-take grazing	Under this system, the pastures are grazed continuously for the whole of the grazing season or for the whole year by a fixed number of 'tester' animals plus a variable number of 'put and take' animals. The latter are varied to graze the pasture at a constant pre-determined stocking pressure.
Pyranometer	The instrument that is used to measure total incoming radiation is called pyranometer.
Pyregeometer	An instrument that measures the effective terrestrial radiation.
Pyreheliometer	An instrument used for measuring the intensity of direct solar radiation at normal incidence.
Pyrolysis	It is a thermal degradation of biomass into volatile and non-volatile products prior to combustion, *e.g.,* when wood is heated in the absence of O_2 it breaks down into many components such as Co, methanol, tar and charcoal etc. This process is also called destructive distillation or weed distillation in short.
Q10	The number of times the rate of process (physical, chemical or physiological) increases with an 10°C rise in temperature. If the rate of the process is doubled Q^{10} is 2. Q^{10} for respiration is 2 to 3.
Quadrat	A sampling unit which has an area of definite size, which may be circular, rectangular or square in shape.

Term	Terminology
Quadratic response	The output goes on increasing up to a certain level of input and then starts decreasing when input is increased further. In such cases the response is said to be quadratic and is given by: $Y = a + bx + cx2$
Quadruple cropping	Growing 4 crops a year in sequence.
Quality seed	Good-quality seeds, generally true to species or variety, have capacity for high germination, are free from diseases, insects, weed seeds, inert matter, other crop seeds and extraneous material.
Quantasomes	Photosynthetic units present in the unit membrane of the granum disc.
Quantum number	The number of molecules of quanta necessary to reduce 1 molecule of CO_2. His also called as the quantum requirement.
Quantum or photon	Terms often used interchangeably for the particles of energy in electromagnetic radiation. (The quantum is the unit quantity of energy in the quantum theory, while the photon is a quantum of the electromagnetic field.)
Quantum yield	Refers to the number of molecules of CO_2 reduced per molecule of quanta absorbed.
Quartiles	Those values which separate a data set into 4 equal parts.
Quillings	Smaller and broken pieces of quills are called quillings. These are used for grinding and also for the distillation of the cinnamon bark oil.
Quills	The long compound rolls of bark (cinnamon) up to 1 m in length which constitutes the best grade.
R value	It is the percentage of the crop land actually cropped in a year. Frequencies of cropping in a fallow cycle (R) is equal to crop year times 100. R values are synonymous to cropping index.
Rabi (winter) crops	Crops which are grown during the winter season (October-November to March-April) with irrigation or on conserved soil moisture, *e.g.,* wheat, barley, oat, chickpea, safflower etc.
Rabi (winter) season	Season during which winter crops like wheat, barley, mustard, gram, linseed, etc. are cultivated and usually it extends from September to April.

Term	Terminology
Raceme	Flower clusters with separate flowers on short equal stalks springing from a main central stem, the lowest flowers opening first.
Radiant flux	The amount of radiant energy emitted, transferred or received per unit time.
Radiation	A term originally signifying the propagation of electro-magnetic radiation through space, but extended to include corpuscular radiation, such as alpha particles, beta particles, neutrons and electrons.
Radicle	The rudimentary root of the embryo which forms the primary root of young seedling.
	The part of the plant embryo that develops into the primary root.
Radio autography	A technique for identification of radioactive compounds on a chromatogram. This is done by exposing the chromatogram to sensitized photographic film which will then develop spots where radio- active substances are present.
Radioactivity	Spontaneous nuclear transformation, the energy of the process being emitted in the form of alpha-, beta- or gamma-rays.
Radioisotope	An isotope of an element exhibiting radioactivity, *e.g.*, pH32, C14 acid, used in tracer studies.
Rain shadow	A comparatively dry area on the leeward side of the high land which stands in the path of rain bearing clouds.
Rainfed dryland	Areas that depend on rainfall for crop production, but are not flooded. The run-off and infiltration of rain water is so high that water does not accumulate on the land.
Rainfed farming	Growing of field crops entirely with rain water received during the crop season (rainfall usually > 800 mm) under humid and subhumid climates and the crops may face little or no moisture stress during their life-cycle.

Term	Terminology
Rainfed wetland	Areas where rainfed rice is grown on puddle soil in fields bounded by dikes capable of ponding water to about 30 cm. Water depth seldom exceeds 30 cm. These areas receive no irrigation and non-rice crops may be grown before or after the wetland rice.
Raingauge	Instrument for measuring the depth of water from precipitation supposed to be distributed over a horizontal impervious surface and not subject to evaporation.
Rainy day	A recorded rainfall of 2.5 mm or more within a period of 24 hours.
Rainy season	That period of the year in which monthly rainfall exceeds 10% of the monthly potential evapo-transpiration (in India May-June to September-October).
Ranching	Grazing of cattle extensively on a fully commercial basis, controlled both by herding and fencing of territories, environmental pressure adequate range quality.
Random sample	A sample for which every individual of the population has an equal chance of being chosen.
Randomization	An act of assigning the treatments to different plots by a random process.
Randomized block design (RBD)	In this design the land on which the trial is to be carried out is divided into as many blocks of the same size and shape as there are replications and each of the blocks into as many plots of the same size and shape as there are treatments.
Range (nuclear)	The thickness of an absorbing material required to remove or absorb all detectable radiation of a particular type.
Range (pasture)	An extensive area of natural pasture land. If unfenced it is an open range.
Range (statistics)	The range of the distribution is the difference between the largest and the smallest of observations and gives some idea of the amount of variability present.

Term	Terminology
Range management	Management of a range to produce maximum forage for sustained use without jeopardising the other resources or uses of land.
Rangeland rejuvenation	Accomplishment of favourable environment to forage plants by soil-plant-animal management practices.
Rate	The amount of active ingredient or acid equivalent applied per unit area or other treatment unit. Rates of formulation per area should not be used in scientific publications.
Ration	A 24-hour allowance of a feed or a mixture of feedstuffs making up the animal diet.
Ratoon	The emergence of new tillers from the root/crown of a plant, specially as in sugarcane, sorghum etc.
Ratoon cropping	The cultivation of an additional crop from the regrowth of stubbles of previous main crop after its harvest, thereby avoiding reseeding or replanting such as in sugarcane, sorghum, rice, fodder grasses etc.
	Cultivation of crop regrowth after harvest, although not necessarily for grain.
Reaper	A machine is used to cut grain crops.
Recession curve	It is the relationship between the time at which water has completely receded from the soil surface and distance along the run in a surface irrigation system.
Recession farming (diara land farming)	It is a system in which crops are planted in flooded areas as the rainy season ends and water recedes. This system takes advantage of thoroughly saturated soil profile and also has the advantage of silt and nutrients left behind by flood water.
Recession phase (irrigation)	Portion of total irrigation time between the beginning of recession at the upper end and the disappearance of the water from the field surface. Time elapsed during this phase is known as recession time.
Recharge	Replenishment of groundwater storage from precipitation, infiltration from surface streams and other sources.

Term	Terminology
Reclaimation (land)	The operation or process of changing the condition or characteristics of land which cannot be utilized to its full potential otherwise, *viz.*, irrigation of land, drainage of swamp or waterlogged land, gypsum application to alkali soils, liming of acid soils etc.
Reclaimation (saline soils)	The process of removing excess soluble salts or excess exchangeable sodium from soils.
Red Data Book	A compilation of data on species threatened with extinction maintained by IUCN.
Red soils	They are formed mainly from granite rocks and are highly variable in depth and show a well-developed profile character. They are coarser grained than the black soils. The percentage of clay increases markedly from the top soil to the subsurface soil. The subsoil at times may be compacted and may not be very permeable to water. The soils grade from the poor, thin gravelly and light-coloured soils of the uplands to the more fertile, deep dark varieties, poor in nitrogen, phosphorus and humus but fairly rich in potassium. Murrum is the gravelly disintegrated rock found beneath the red soils.
Redox potential (oxidation-reduction potential, Eh)	A measure of the tendency of a given oxidation-reduction system to donate electrons or to accept electrons. The Eh is determined by measuring the electrical potential difference in volts or millivolts between the given system and a standard redox system.
Reduced sample	A representative part of the aggregate sample obtained by a process of reduction in such a manner that the mass approximates to that of the final (laboratory) sample.
Reduced tillage	A tillage system in which the primary tillage operation is performed in conjunction with special planting procedure in order to reduce or eliminate secondary tillage operations.
Reducing and non-reducing sugars	Sugars possessing a free aldehyde or ketone group are reducing sugars, such as maltose. When the above groups are locked up in a glycosidic linkages in the formation of complex sugars, the resulting sugars are said to be non-reducing, *e.g.*, sucrose.

Term	Terminology
Reference crop evapo-transpiration (ETo)	Rate of evapotranspiration from an extended surface of 8 to 15 cm tall, green grass cover of uniform height, actively growing, completely shading the ground and not short of water; mm/day.
Reflection coefficient	Expresses the reflectivity of the total short-wave radiation while albedo is used exclusively for the visible light.
Refractive index	It is the ratio of the velocity of light in a vacuum to its velocity in the substance.
Regeneration	Sprouting of stubbles following cutting and/or grazing; synonymous to regrowth.
Registered seed	It is the progeny of breeder seed or foundation seed handled by a producer acceptable to the certifying agency to maintain satisfactory genetic purity and identity.
Registration	The process designated by FIRA and carried out by EPA by which a pesticide is largely approved for use.
Regression	It is a quantitative measure of relationship between 2 variables which helps to predict the value of one variable (dependent) from the known value of the other (independent) variable.
Regression coefficient	A numerical measure of the rate of change of the dependent on the independent variable.
Regression line	Straight line which represents the relationship between 2 variables.
Regur (soil)	Dark gray or black soil often rich in calcium carbonate (0 to 30%), sticky when wet, cracks in drought, pH varies from 6.5 to 9.0. It is a grumosol, *i.e.* black cotton soil.
Reichert-Maeissel value	It is the R-M value of a fat and is the number of milliliters of 0.1 m potassium hydroxide required to neutralise the fatty acids volatile in steam, in 5 g of fat.
Relative crop-intensity	It determines the amount of area-time allocated to 1 crop or a group of crops relative to the area-time actually used in production of all crops.

Term	Terminology
Relative crowding coefficient (k)	It is a measure of the relative dominance of one component crop over the other in an intercropping or mixed cropping system. The coefficient (k) is determined separately for each component crop.
Relative density (RD)	The population of a particular species expressed as a percentage of all the species in a habitat. R.D. = Density of a particular species ÷ Total density of all the species × 100.
Relative dominance (RO)	The actual area occupied by a species expresed as a percentage of total area occupied by all species in a given land area.
Relative growth rate (RGR)	An index of the amount of growing material per unit dry weight of plant per unit time.
Relative humidity (RM)	A measure of moisture content of air expressed as percentage. It is the ratio of the mass of water vapour in a given volume of air at a given temperature to the maximum quantity of water vapour which the same volume of air could hold at the same temperature.
Relative net return (intercropping)	It is the ratio between the monetary value of base and intercrop plus differential cost of cultivation (intercropping-monoculture of base crop) and monetary value of base crop in monoculture
Relative spread index	It represents an area of the crop expressed as percentage of the total cultivated area in the zone divided by the area of the crop expressed as percentage of the total cultivated area in the country, times hundred.
Relative turgidity (RT)	Ratio of turgidity of leaf tissue at a particular stage to that of when it is fully turgid.
Relative weed	These are crop plants which when grow at the places where these are undesirables.
Relative yield index	It represents the mean yield for a crop in a region divided by the mean all India yield of the same crop times hundred.
Relative yield total	This implies that the mean of plant relative yield (ratio of the per plant yields in a mixture) to that in monoculture will have a value close to unity. In terms of relative yields, the value is a total, which has been called the relative yield total.

Term	Terminology
Relay cropping	Seeding or planting of succeeding crop after flowering and before the harvest of the standing crop. It is analogous to a relay race where 1 crop hands over the land to the next crop in quick succession, *e.g.,* sowing toria in standing crop of maize.
Relay intercropping	Growing two or more crops simultaneously during part of the life cycle of each. A second crop is planted after the first crop has reached its reproductive stage but before it is ready for harvest.
Relief	The configuration of the land or the vertical variation in terrain including such features as direction and degree of slope, sharpness of change in slope direction and variation of these features.
Remote sensing	Acquisition of physical data of an object without touch or contact.
Renewable sources of energy or non-conventional energy	Energy from biomass, *i.e.,* from fuel wood, agricultural and forest wastes and animal wastes, constitutes renewable sources of energy as these are primarily obtained by harnessing the solar energy which is inexhaustible.
Replicate (replication)	Multiple repetition of a treatment within a trial or experiment.
Research	Studious inquiry involving critical and exhaustive investigation, including experimentation, designed to discover new facts.
Research-managed trials	Experiments done in farmer's fields, but managed by researchers to attain higher degree of experimental precision while still getting the effect of variables existing on the farms.
Reseeding cultivar	It perpetuates itself by volunteering from shattered seed; usually made possible because of a high percentage of hard seed or seed with high dormancy.
Residual effect	Effect of the previous crop in a sequential cropping pattern on the productivity of the current crop.
Residual effect of manure (fertilizer)	This refers to the residual beneficial effect of application of organic manure or fertilizers on the succeeding crops due to unutilized plant nutrients left-over by the preceding crop.

Term	Terminology
Residue (pesticides)	That quantity of a substance, especially of active pesticide, remaining in soil or plant.
Resilience (crop)	The resilience of a crop is the extent of its ability to survive a perturbation or stress with minimum effect on yield. It is, thus, critical component of the crop-yield stability. In a sole crop, recovery after a shock may involve compensation within and between plants of similar genotypes. In an intercrop, compensation is often also between components.
Resistance	The degree to which a species of plant or other organism tolerates a toxic substance.
	The ability of a plant species to withstand the phytotoxicity of a chemical. Herbicide resistance, in the context of weeds, is defined as a characteristic of the weed species to withstand an herbicide dosage substantially higher than the wild type of the same plant species can withstand.
Resource-use indices (RUI)	It is defined as the amount of an environment resource used in different weed-control treatments (crop + associated weeds) divided by the amount of the same resource used by manual weeding, *i.e.,* weed-free crop at harvest.
Respiration	The process by which energy is acquired in a living plant organism, by the breakdown of complex organic molecules, with the release of waste carbon dioxide and water.
Respiratory quotient (RQ)	The ratio of the volume of CO_2 released to the volume of O_2 absorbed in the respiratory process ($RQ = CO_2/O_2$), It indicates the type of substrate oxidized. If carbohydrates are substrates, the ratio is 1. For fats it is less than 1 and for organic acids it is more than 1.
Response curves	A close functional relation between the input x (nutrients) and output Y (production).
Restoration grazing	Intensive system of management whereby grazing is deferred on various parts of the range during succeeding years, allowing, the defended part complete rest for 1 year; 2 or more units are required.

Term	Terminology
Restorative crops	Such crops provide a good harvest along with enrichment or amelioration of soil.
Restorer line	An inbred line that when crossed on a male-sterile strain, causes the resulting hybrid to be a male fertile and produce pollen.
Result demonstration	A type of demonstration in which the results of proven technology are shown in comparison to existing practices.
Retting	Soaking flax, hemp, jute or other fibrous plant material in water to facilitate separation of the fibers from the stalk for further processing.
Reverse metabolism	Breakdown of nontoxic herbicides' inside the plant into toxic metabolites is known as reverse metabolism.
Reverted phosphoric acid	The part of the water-soluble P_2O_5 in a fertilizer which as a result of some reaction has become insoluble in water.
Rhizobia	A group of symbiotic bacteria capable of fixing atmospheric nitrogen in the root nodules of legumes from which they receive their energy.
Rhizome	Underground stem, usually horizontal and capable of producing new shoots and roots at the nodes.
	The horizontal, slender, underground root-like stem capable of sending out roots and leafy shoots.
Rhizosphere	That portion of the soil in the immediate vicinity of plant roots in which the abundance and composition of the microbial population are influenced by the presence of roots.
Ribonucleic Acid (RNA)	Polymeric constituent of all living cells, consisting of a single strand of alternating phosphate and ribolleunitswith the bases adenine, guanine, cytosine and 'uracil bonded to the ribose, the structure and base sequence of which are determinants of protein synthesis.
Ribose nucleic acid (RNA)	A nucleic adds which controls the sequence of amino-acids in an enzyme or other proteins. The synthesis of RNA is controlled by the DNA of a gene.

Term	Terminology
Rice glazing	Polished rice treated with talc (and in some cases with glucose).
Rice milling	It involves the removal of hulls and bran from husked rice. Milling ranges from hand pounding to modem milling equipments.
Ridge former (ridger)	An implement used to form ridges and furrows or channels.
Ridge planting	A method of planting crops on ridge formed through tillage operations.
Ridge terrace	A long, low ridge of earth with gently sloping sides and shallow channel along the upper side, to control erosion by diverting surface run-off across the slope instead of permitting it to flow uninterrupted down to slope.
Rill erosion	Erosion producing small channels that can be obliterated by tillage.
Ring lines (area)	The gross plot area of treatment minus (–) net plot area is called ring area or ring lines, having uniform planting rows in both directions in field experiments.
Riparian crops	Grown along irrigation and drainage channels or waste bodies.
Ripening stage	Period between completion of grain formation and maturity.
Rock phosphate	Rock containing phosphorus such as tricalclum phosphate. Rock phosphate is treated with strong adds to make superphosphate. Finely ground tock phosphate is used in strongly add soils to supply P_2O_5. The finer the rock phosphate more effective it will be on add soils.
Rogue	An off-type plant in a crop field.
Root crop	Crop plants whose surplus or reserve foods are stored primarily in enlarged roots, *e.g.* sweet potato.
Root nodules	Refers to small swelling on roots of leguminous plants produced as a result of infection by nitrogen-fixing bacteria (rhizobia).
Root sucker	A shoot arising from the root of a plant.
Root surface sorption zone	The soil volume immediately around the roots.

Term	Terminology
Rosette	In many plants profuse leaf development with retarded internodes growth to give bunchy or rosette appearance.
	A circular cluster of leaves.
Rotary tillage	Tillage operation employing rotary action to cut break and mix soil.
Rotation	Repetitive cultivation of an ordered succession of crops (or crops and fallow) on the same land. One cycle often takes several years to complete.
Rotation pasture	A pasture used for a few seasons and then ploughed for other crops.
Rotational grazing	Dividing a large pasture into several small paddocks and allowing the cattle to graze each, periodically perhaps every 4 weeks.
Rough rice (paddy)	Unhulled rice.
Roughage	Plant materials (feedstuffs) that are relatively high in crude fiber and low in digestible nutrients such as straw and stover.
Roughing	To remove weeds or off-type or diseased plants from a standing field crop.
Row cropping	Growing 2 or more crops simultaneously where 1 or more crops are planted rows.
Row crops	Crops which are sown in rows keeping uniform spacing throughout the field, *e.g.,* cotton.
Row intercropping	Growing two or more crops simultaneously where one or more crops are planted in rows.
Row marker	An attachment to a sowing machine to mark a line on the field to, guide the position of extreme furrow opener during the next trip.
Rubble land	Land area with 90% or more of the surface covered with stones and boulders called rubble land.
Runner	Trailing stems or branches that take root and reproduce new plants at the joints or ends.
Run-off	The portion of the precipitation which is not absorbed by the soil but finds its way into the streams after meeting the persisting demands of evapotranspiration interception and other losses.

Term	Terminology
Run-off farming	It refers to the farming where harvested water is stored in the soil itself.
Safener	Substance capable of antagonizing specific herbicide phytotoxicity to plant (also called Antidote).
Saline alkali soil	A soil containing a combination of soluble salts and exchangeable sodium sufficient to interfere with the growth of most crop plants. The electrical conductivity and exchangeable sodium of the saturation extract are more than 4 millimhos/cm at 25°C and more than 15% respectively. The pH is usually 8.5 or less in the saturated soil paste.
Saline soil	A soil containing excess of neutral soluble salts to impair soil and crop productivity and has pH value of 8.5, ECe 4 mmhos/cm at 25°C and ESP 15%.
Salinization	The process of accumulation of soluble salts in a soil usually chlorides and sulphates.
Salt	The product other than water of the reaction of a base with an acid.
Salt balance	The relation between the quantities of dissolved salts carried to an area in irrigation water and the quantity of dissolved salts removed by drainage water.
Salt index	Concerning fertilizer salts and compound fertilizers; an index of the extent to which a given amount of fertilizer increases the osmotic pressure of soil solution.
Saltation	Soil particle movement in water or wind where particles skip or bounce along the stream bed or soil surface.
Sample	A relatively small group of individuals taken from a very large population about which we seek information.
Sampling	The procedure of selection of a sample from population is known is sampling.
Sampling error	Deviation of a sample value from the true value owing to the limited size of sample.
Sampling unit	A defined quantity of material having a boundary which may be physical, *e.g.*, a container, or hypothetical *e.g.*, a particular 1ittle time interval in the case of a flow of material.

Term	Terminology
Sand	Small rock 01 mineral fragments having diameters ranging from 0.05 to 2.0 mm.
Sandy clay	Soil of this textural class contains 35% or more of clay and 45% or more of sand.
Sandy clay loam	This class of soil contains 20-35% clay, less than 28% silt and 45% or more of sand.
Sandy loam	Soil of the sandy loam class of texture which has 50% sand and less than 20% clay.
Sandy soil	A broad term for soils of the sand and loamy sand classes where more than 70% sand and less than 15% clay are present
Sapling	Tree (plant) seedlings, young tree plants having age of 6 months to about 1 year of forest trees or plantation crops.
Saponification value	It is the number of milligrams of potassium hydroxide required to neutralize the fatty add resulting from the complete hydrolysis of 1 g of the substance.
Saprophytic	Obtaining nourishment from non-living biological materials.
Satellite weed	Weed that has become an integral part of a crop ecosystem.
Saturated air	Moist air in a state of equilibrium with a plane surface of pure water or ice at the same temperature and pressure. In such a state the relative humidity is 100% and the amount of water vapour is maximum for the given temperature.
Saturated fatty acids	Fatty acids in which the carbon bonds are fully saturated, *i.e.,* those fatty acids where carbon to carbon bonding is not more than 1.
Saturated soil	A soil which has its interspaces or void spaces completely filled with water to the point where run-off occurs.
Saturated vapour pressure	The vapour pressure of a parcel of saturated air at a given temperature.

Term	Terminology
Saturation deficit	The difference between the actual vapour pressure and saturation vapour pressure at the existing temperature. The additional amount of water vapour needed to produce saturation at the current temperature and pressure, expressed in grams per cubic meter. Also known as vapour pressure deficit.
Saturation extract	The solution obtained by pressure or vacuum filtration of a soil paste that has been made up to a saturated condition by adding water while stirring.
Saturation index	An estimate of carbonate precipitation from irrigation water as a function of the degree of calcium carbonate saturation of the son solution.
Saturation percentage	The moisture percentage of saturated soil expressed on a dry-weight basis.
Savanna	Grassland with scattered trees either as individuals or clumps often a transitional type between true grassland and forest.
Scarification (seed)	Scratching of hard seed-coat to remove the mechanical barrier to obtain uniform germination with better penetration of moisture into the seed.
Scientific name	The binomial Latin name of a species consisting of the generic name, a Latin or latinized noun, followed by the species epithet a Latin or latinized adjective.
Scintillation	A flash of light produced in a phosphor by radiation.
Scintillation counter (nuclear)	A counter employing a phosphor, photomultiplier tube and associated circuits for the detection of radiation.
Sciophytes	Shade-tolerant species which do not require more light.
Scouring	The self-cleaning flow of soil over tillage tools through a sliding action.
Scouring or pearling (rice)	Process of bran removal (locally known as polishing).
Seasonal consumptive use	The total amount of water consumed in evaporation and transpiration by a crop during the entire growing season, expressed in depth or volume of water per hectare.

Term	Terminology
Seasonal drought	A kind of drought which occurs during distinct annual periods of dry weather.
Secondary compounds (plant)	The compounds produced-in the plant body as by-products of primary metabolic pathways, *e.g.,* glucosinolates in crucifers, hydrocyanic acid (HCN) in cassava and sorghum.
Secondary fertilizer elements	Calcium, magnesium and sulphur which are next in importance to major nutrients for optimum crop growth.
Secondary growth	Growth which results from the activity of vascular or cork cambium in the formation of secondary tissues and mainly resulting in an increase in diameter or stem girth of a plant or tree.
	Growth resulting from the formation of new cells by cambium.
Secondary metabolites	A group of metabolites including toxic substances or waste substances, possibly sometimes serving as reserves or just accidental end-products, such as alkaloids, terpenes, glycosides and tannins.
Secondary mineral	A mineral resulting from the decomposition of a primary mineral or from the reprecipitation of the products of decomposition of a primary mineral.
Secondary plant nutrients	The secondary plant nutrients are calcium, magnesium and sulphur and are called secondary because these are not applied as straight commercial fertilizers.
Secondary productivity	The productivity of pasture or grassland, measured in terms of animal gains or products such as milk, meat or wool instead of herbage yield, is called as secondary productivity. This is an indirect measurement of pasture productivity.
Secondary tillage	The tillage operations following primary tillage to create a good seed-bed fur proper seeding or planting.
Sedge	Plant belonging to the family cyperaceous, a large family of monocotyledonous plants distinguished chiefly by having achiness, solid stems and three ranked stem leaves.

Term	Terminology
Seed	Sexually or vegetative propagated planting materials which are used for seeding and planting and as such should be free from (pests and diseases) any infection and should give a good crop stand by good seeding.
Seed certification	A means to maintain and make available to the public, sources of high quality seeds and prorogating materials of superior varieties so grown and distributed as to insure genetic identity. This is done by means of inspections of fields and seeds and by regulations for checking on the production, harvesting and cleaning of each lot of seed.
Seed control	Seed law enforcement which applies to all seeds sold commercially. Correct labeling with respect to the kind of seed, germination, purity and other quality factors are ensured.
Seed dressing	Chemical treatment of seeds before sowing to control any disease or pest attack in growing crops, particularly cereal.
Seed drill	A machine to place seeds at uniform rate at selected depth and row spacing.
Seed hardening	It is a process of subjecting seeds before sowing to alternate cycle of wetting and drying to induce tolerance to drought.
Seed rate	The quantity of seed required for sowing in a unit area of land.
Seed treatment	The process of application of a thin coat of insecticide or fungicide over seeds or similar applications to prevent pest infestation.
Seed-bed	A well-prepared land for sowing or planting.
Seed-cum-fertilizer drill	An implement which drills seed and fertilizer simultaneously and uniformly in the same or different rows.
Seeder	A device for placing seeds on or in the soil.
Seedling	The juvenile stage of a plant grown from seed. Usually indicates plants which have up to and including about 4 true leaves.
Seedling stage	It includes the period from emergence till just before the appearance of first tiller.

Term	Terminology
Seepage	The slow movement of water through small cracks, pores, interstices, etc., in the surface of unsaturated material into or out of a body of surface of subsurface water.
Selection	In a population the preservation of certain individuals that have desirable characteristics.
Selective herbicides	A chemical that is more toxic to some plan species than to others.
Selectivity	Property of differential tolerance; essential for the proper functioning of herbicides that kill some plant species when applied to a mixed population, but without serious damage to other species.
Selectivity index	It is the ratio of maximum dose of herbicide tolerated by crop to minimum dose of herbicide required for weed control.
Self-absorption (nuclear)	The absorption of radiation by the radioactive substance itself. Self-fertilization, Fertilization of an egg by a pollen grain from the same plant.
Self-mulching soil	A soil in which the surface layer becomes so well aggregated that it does not crust and seal under the impact of rain but instead serves as surface mulch upon drying.
Self-pollination	Transfer of pollen from the anther to the stigma of the same flower or another flower on the same plant.
Semi-arid	Describes a land or climate where the rainfall is greater than 750 mm. The PET is more than precipitation. The moisture deficit and PE indices range from -33.3 to -66.6 and 16 to 31, respectively.
Semi-arid zones	These are zones delineated by Thomthwaite's moisture indices varying from -20 to –40. These indices are defined by 100 S -60 D/N, where S = mean annual water surplus, 0= mean annual water deficiency and N = mean annual water need.
Semi-dwarf wheat	These are short-statured wheat, having sturdy straw, strong crown roots, more fertile florets, higher tillering capacity, earliness, higher harvest index, disease resistance and are high-yielding, fertilizer-responsive, photo-insensitive and have wide adaptation.

Term	Terminology
Seminal root	A root arising from the base of hypocotyls, as in wheat.
Semi-organic fertilizer	Product in which declared nutrients are of both organic and inorganic origin obtained by mixing and/or chemical combination of organic and inorganic (mineral) fertilizers.
Semi-permeable membrane	A membrane which permits the solvent but not the solute to pass through it.
Sensor	The component of an instrument that converts an input signal into a quantity that is measured by another part of the instrument.
Sequential cropping	The growing of different crops on a piece of land in a pre-planned succession.
	Growing two or more crops in sequence on the same field per year. The succeeding crop is planted after the preceding crop has been harvested. Crop intensification is only in the time dimension. There is no intercrop competition. Farmers manage only one crop at a time in the same field.
Sesquioxide	An oxide which contains 2 metallic atoms to every 3 of oxygen. Commonly refers to iron and aluminium hydroxides.
Sewage farm	A farm which takes sewage, usually as sludge, from settlement tanks; the sewage is used as manure and the effluent is drawn off to irrigate the land. A farm with a sandy soil is preferred for sewage disposal since sand is porous and acts as a filtering material.
Sexual reproduction	Reproduction resulting from the fusion of gametes (sex cells).
Shade trees	Fast growing trees which provide shade to many crops like coffee, tea, cocoa, and some orchards for their growth and development in tropics.
Sheer erosion	The removal of a fairly uniform layer of soil or material from the land surface by the action of rainfall and run-off water.
Shelling	Removal of grains from cobs or kernels from pods or husk from paddy.

Term	Terminology
Shelter belt (wind break)	A wind barrier of living trees and shrubs established and maintained for protection of crop fields.
Shifting cultivation	The practice of cultivating clearings scattered in the reservoir of natural vegetation (forest or grass woodland) and of abandoning them as soon as the soil is exhausted; and this includes the practice of shifting homesteads in order to follow the cultivator's search for new fertile land.
Shoot	A collective term for a stem and its leaves.
Short-day plants	The plants that normally develop flower with photoperiod (day-light period) of 8-10 hours, *i.e.,* When the photoperiod is less than a certain critical maximum.
Short-wave radiation	Radiation with wavelengths less than 4 microns.
Shrub	Woody perennial plant or tree, low in stature with a habit of branching from the base, with bushy appearance.
Sickle	A curved steel blade having a hand grip, used (or harvesting by manual power. Blade may have smooth or serrated edge.
Side draft	The horizontal component of the pull perpendicular to the direction of motion. This is developed if the centre of resistance is not directly behind the centre of pull.
Side-dressing	The application of commercial fertilizer along the side of a row or around a plant.
Sieving	Separation of desired grains by a mechanical device, where the desired grain penetrates the device and the undesired material is carried over the device.
Sigmoid growth curve	A graph showing the rate of growth of a crop (annuals) for a period of time, Sigmoid means S-shaped. The general pattern is one of initially small increases in size, followed by rapid growth and then by a period during which the size of the crop increases slowly or not at all.
Significance level (statistics)	A probability value that it is considered so rare in the sampling distribution specified under the null hypothesis that one is willing to assert the operation of non-chance factors. Common significance levels are 0.05 and 0.01.

Term	Terminology
Silage	Animal feed resulting from the storage and fermentation of green or wet crops under anaerobic conditions.
Silage additive	Material added to forage at the time of ensiling to enhance either its preservation or feeding value.
Silage crop	Those crops which are preserved in a succulent condition by partial fermentation in a tight receptacle.
Silage preservative	Material added to forage at the time of ensiling to enhance the favourable fermentation process.
Silica sesquioxide ratio	The molecules of silicon dioxide (SiO_2) per molecule of aluminium oxide (Al_2O_5) plus ferric oxide (Fe_2O_3) in clay minerals or in soils.
Silk (corn)	The stigma and style of the female com flower through which the pollen-tube grows to reach the embryo sac.
Silo	A semi-airtight structure designed for use in the production and storage of silage.
Silt	Small mineral soil particles of a diameter of 0.002 - 0.05 mm. Silt loam: Soil material having 30% or more silt and 12-27% of clay or 50-80% of silt and less than 12% of clay.
Silt traps	These are used for collecting silt received through irrigation water. These are placed at the junction of 2 points in the water conveyance line.
Silting	The deposition of water-borne sediments in lakes reservoirs, stream Channels or over flow areas.
Silty-clay loam	Soil having 27-40% of clay and less than 20% of sand.
Silviculture	It is the art and science of cultivating forest crops as distinguished from silvics which deals with the study of the life-history and general Characteristics of forest trees or crops.
Silvopastoral system	It is a land management system in which forests are managed for the production of wood as well as for the rearing of domesticated animals. In this system, animals are kept and permitted to graze within the forest.

Term	Terminology
Simple correlation	It measures how 2 variables (for example yield and spikelet number/plant) are associated in a sample. The correlation coefficient(s) can be regarded is a measure of the intensity of the linear association and the value ranges from 0 to 1.
Simple fruit	A fruit formed from single ovary without other parts adhering to it.
Simple linear regression	Is a mean of evaluating the linear relationship of one independent or predicator variable, X to a single usually continuous, dependent or response variable Y.
Simplest	Functionally integrated unit consisting of all living cells of a multi-cellular plant; all living cells connected by plasmodesmata; including the phloem.
Simpson's index (S.I.)	Simpson's index is a measure of concentration of dominance, can be used to determine the degree of diversity in a given community type.
Simultaneous poly culture	Simultaneous growth of two or more useful plants on the same, plot. It includes mixed cropping, intercropping, interculture, interplanting and relay planting.
Single grain soil	A structure less soil in which each particle exists separately as in sand-dunes.
Single-cross parent	An F_1 offspring of two inbred pare.nts which in turn is used as a parent usually with another single cross parent to produce a double-cross hybrid as in com.
Sink	Term used for storage organs where photosynthates are stored (grain, tubers) or the term usually applied to refer to the plant parts that are involved in utilization and storage of photosynthates, for example grains; fruits etc. The site in a plant with a high rate of metabolic, activity where food resources are used.
Sink capacity	The capacity of the spikelets or reproductive or economic plant part like earhead, cob, tuber etc. to accept photosynthates.
Siphons	These are bent pipes in desired shapes and sizes which are made up of rubber, polythene or aluminium for conveying water from the water channel into irrigation beds or furrows.

Term	Terminology
Skewed distribution	Distribution which departs from symmetry and tails off to one end.
Skip cropping	A line is left unsown in the regular row series of sowing is called skip cropping.
Skip row planting	When a line is left unsown in regular sowing.
Slash and bum system	A kind of shifting cultivation usually in high-rainfall region where the cropping period is followed by a fallow period of several years during which bush or tree growth occurs. For the next cycle of cropping, the bush or tree growth is again cleared by cutting and burning.
Sleet	When rain falls through a cold layer of air near the earth's surface rain drops get frozen into, ice and fall down. This form of precipitation is called sleet.
Slick spots	Wet spots with high exchangeable sodium occurring generally in salt-affected areas.
Slip (prorogation)	A shoot or leaf cutting to be rooted for vegetative propagation.
Slow neutrons	Neutrons possessing a kinetic energy equal to or less than 100 eV.
Slow-release fertilizer	Fertilizer whose nutrients are present as a chemical compound or in a physical state such that their availability to plants is spread over a period of time.
Sludge	The semi-solid part of sewage/water that has been sedimented or acted upon by bacteria.
Slurry	Semi-liquid effluent from livestock, consisting of urine and fasces, possibly diluted with water.
Smother crop	A dense growing crop which can suppress or stop the growth of a less competitive crop; it is grown for the suppression of weeds.
Snow gauge	Apparatus designed to measured the amount of water fallen in the form of snow.
Snow line	The altitude above which snow lies throughout the year.
Snow stake	Graduated fixed stake used in regions of abundant snowfall to facilitate the measure of snow depth.

Term	Terminology
Soaking	The part of presowing seed treatment or parboiling where paddy is soaked in water to raise its moisture content.
Social forestry	It is a programme of forestry development and conservation under various agro-climatic and soil conditions through (i) mixed plantation on waste-lands and panchayat lands and (ii) reforestation of degraded forests and raising shelter belts.
Sod	Top few centimeters of soil permeated by arid held together with grass roots or grass-legume roots.
Sod culture	A system of soil management wherein the plants are grown in permanent grass Without tillage.
Sod seeding	Mechanically placing seed, usually legumes, small grains, directly into a grass sod.
Sod-bound	Used to describe the unproductiveness of grass sod due to lack of nitrogen.
Sodium adsorption ratio (SAR)	A ratio for soil extracts and irrigation waters used to express the relative activity of sodium ions in exchange reactions with soil: $$SAR = \frac{Na^+}{\sqrt{Ca^{++} + Mg^{++}}/2}$$
Soft water	Water which does not contain those minerals that prevent free lathering when soap is added.
Softener	A substance which reduces toxicity on a specific herbicide to the crop.
Soil	A collection of natural bodies developed in the unconsolidated mineral and organic material on the immediate surface of the earth that serves as a natural medium for the growth of land plants and has properties· due to the effects of climate and living matter acting upon parent material, as conditioned by topography, over a period of time.
Soil (mature)	A soil which has reached the full development to be expected under existing weathering and biological processes.
Soil adhesion	The sticking of soil to tillage tools or wheels.

Term	Terminology
Soil aeration	The exchange of air in soil with atmospheric air.
Soil aggregate	A naturally occurring cluster or group of soil particles in which the forces holding the particles together are much stronger than the forces between adjacent aggregates.
Soil air	Air occurring in soil pores; the gaseous phase of the rest, being that volume not occupied by solid or liquid.
Soil alkalinity	The degree or intensity of alkalinity of a soil expressed by a value 7.0 on the pH scale.
Soil amendment	Any substance that is added to the soil for the purpose of improving its physical or chemical character, enhancing soil productivity or promoting the growth of crops but excluding commercial fertilizers and organic manure.
Soil application	Chemical application made primarily to the soil surface rather than to vegetation.
Soil association	A group of defined and named taxonomic soil units occurring together in an individual and characteristic pattern over a geographic region comparable to plant associations in many ways.
Soil auger	A tool for boring into the soil and withdrawing a small sample of soil for laboratory analysis.
Soil colloids	The colloidal fraction of the soil primarily made up of inorganic material (clay) with varying amounts of organic colloids (humus).
Soil compaction	Reduction in the specific volume of soil by means of mechanical manipulation.
Soil complex	A mapping unit used in detailed soil surveys where 2 or more defined taxonomic units are so intimately intermixed geographically that it is undesirable or impractical, because of the scale being used, to separate them. A more intimate mixing of smaller areas of individual taxonomic units than that described under soil association.
Soil conditioners	The chemicals which are added to maintain physical condition of the soil, *e.g.,* polyvinylites, polyacrylates, cellulose gums, lignin derivatives and silicates.

Term	Terminology
Soil conservation	The preservation of soil against deterioration and loss by using it within its capabilities and applying the conservation practices needed for its protection and improvement.
	A combination of all management and land use methods which safeguard the soil against depletion or deterioration by natural or by man-induced factors.
Soil cultivation	Tillage operations performed to create soil conditions conducive to improve aeration, infiltration, compaction and moisture conservation as well as to control weeds.
Soil cutting	Soil separation from the soil mass by slicing action. Soil degradation. It is a process of changing a soil from one type to another more highly leached one; particularly bringing about replacement of sodium by hydrogen by leaching a saline or alkali soil.
Soil evaporimeter	Evaporimeter used to measure the amount of water evaporated from the ground surface during a given time interval.
Soil extract	The solution separated from soil suspension or a soil at particular moisture content.
Soil failure	The alteration or destruction of a soil structure caused by mechanical forces such as shear, compression and impact.
Soil fertility	Soil fertility refers to inherent capacity of a soil to supply nutrients to plants in adequate amount and in suitable proportion.
Soil genesis	The mode of origin of the soil with special reference to the processes responsible for the development of the solum or true soil from the unconsolidated parent material.
Soil heaving	The swelling of soil resulting from natural forces, such as freezing.
Soil incorporation	Mechanical mixing of any crop residue and agricultural chemical into the soil.
	Mechanical mixing of herbicides in the soil.

Term	Terminology
Soil injection	Placement of the herbicide beneath the soil surface with a mixing or stirring of the soil as with an injection blade, knife or time larger field.
Soil management	The sum total of all tillage operations, cropping practices, fertilizer, lime and other treatments, conducted on or applied to a soil for the production of plants.
Soil monolith	A vertical section of a soil profile removed from the soil and mounted for display or study.
Soil morphology	A systematic study and description of the characteristics of the soil profile in the field.
Soil organic matter	The organic fraction of the soil that includes plant and animal residues of various stages of decomposition, cells and tissues of soil organisms, and substances synthesized by the soil population.
Soil permeability	The quality or state of a soil or any horizon in the soil profile relating to the transmission of water or air to all parts of the mass.
Soil persistence	Refers to the length of time that an herbicide remains effective in the soil and exhibits some degree of phytotoxicity to some plant species.
Soil persistence (herbicide)	Refers to a period of time that a herbicide application on or in soil remains effective.
Soil porosity	The percentage of the soil volume not occupied by solid particles, including all pore space filled with air and water.
Soil productivity	It is the present capacity of a soil to produce crop yield under a defined set of management practices. It is measured in terms of the yield in relation to the input of production factors.
Soil profile	A vertical section of the soil from the surface through all its horizons into the parent material.
Soil reaction	The acid or alkaline reaction of a soil as measured by one or number of methods expressed in terms of pH units, generally in the range of 4.5 to 9.5 pH.
Soil salinity	The amount of soluble salts in a soil, expressed in terms of percentage, parts per million (ppm) or other convenient ratios.

Term	Terminology
Soil science	That science dealing with soils as a natural resource on the surface of the earth including soil formation, classification and mapping, and the physical, chemical, biological, and fertility properties of soils per se; and these properties in relation to their management for crop production.
Soil separates	Individual soil particles after standard dispersion treatment recognized by the International Society of Soil Science are (a) coarse sand, 2.0 to 0.2 mm; (b) fine sand, 0.2 to 0.02 mm; (c) silt, 0.02 to 0.002 mm; and (d) clays, 0.002 mm diameter.
Soil shattering or pulverization	The general fragmentation of a soil mass resulting from the action of tillage forces.
Soil sickness	Condition of soil when the productivity is reduced in spite of available nutrients.
Soil solaraization	The process of heating the surface soil by raising the soil temperatures
Soil specific gravity (Asg)	Ratio of the weight of water-free soil to its volume; also called bulk density; g/cm^3.
Soil sterilant	An herbicide that prevents the growth of plants when applied to the soil. Soil sterilization effects may be temporary or relatively permanent. Herbicide which renders the soil incapable, of supporting growth of all plants; sterilization may be temporary (a few months) or relatively permanent (years).
Soil structure	The combination or arrangement of primary soil particles into secondary particles, or groups of aggregates.
Soil texture	An expression of the distribution of the various particle sizes present in soils. A soil is described as a coarse, medium or fine textured depending on the predominant particle size.
Soil thermometer (geothermometer)	Thermometer for measuring the temperature of the soil at various depths. It is generally mercury-in-glass type thermometer with stem bent at 90" or 120° and scale facing upwards.

Term	Terminology
Soil type	A subdivision of a soil series based mainly upon the texture of the upper layers.
Soil water potential (capillary)	The amount of energy that must be expended per unit quantity of pure water in order to transport reversibly and isothermally an infinitesimal quantity of water, identical in composition to the soil water, from a pool at the elevation and the external gas pressure at the point under consideration to the soil water.
Soil water potential (gas pressure)	This potential component is to be considered only when external gas pressure differs from atmospheric pressures as in a pressure membrane apparatus.
Soil water potential (gravitational)	The amount of energy that must be expended per unit quantity of pure water in order to transport reversibly and isothermally an infinitesimal quantity of water identical in composition to the soil water, from a pool at a specified elevation and atmospheric pressure, to a similar pool at the elevation of the point under consideration.
Soil water potential (osmotic)	The amount of energy that must be expended per unit quantity of pure water in order to transport reversibly and isothermally an infinitesimal quantity of water from a pool of pure water, at a specified elevation and at atmospheric pressure, to a pool of water identical in-composition to the soil water, but in all other respects being identical to the reference pool.
Soil water potential (total)	The amount of energy that must be expended per unit quantity of pure water in order to transport reversibly and isothermally an infinitesimal quantity of water from a pool of pure water at a specified elevation and at atmospheric pressure to the soil water. Consists of osmotic, gravitational and capillary potential.
Soilage	Forage crops cut green and fed to livestock while in the fresh condition are known as soiling crops or as soilage.
Soil-eradibility factor	The soil erodibility factor indicates the inherent erodibility of a soil. It gives an indication of the soil loss from a unit plot, 22 mm long with a 9% slope and continuous fallow culture.

Term	Terminology
Soiling crops	Grown to harvest while they are still green and fed fresh to livestock in stalls.
Soil-moisture characteristic curve (moisture retentive curve)	A graph showing the relationship between the amounts of water remaining in the soil at equilibrium as a function of matric suction.
Soil-moisture deficit	The amount of water needed to return the soil moisture status to the field capacity.
Soil-moisture stress	It is the sum of soil-moisture tension and the osmotic pressure of the soil solution.
Soil-moisture tension	It is the force with which the water is held by soil, against gravity. It is generally expressed in atmospheres or in mm of mercury.
Soil-saving dam	A dam of earth or other material placed across a gully or other water course to impound silt and surface run-off and to control erosion.
Soil-water stress	Sum of soil water tension and osmotic pressure to which water must be subjected, to be in equilibrium with soil water; also called soil water potential; atmosphere or bar.
Soil-water tension	Force at which water is held by soil or negative pressure or suction that must be applied to bring water in a porous cup into static equilibrium with the water in soil. Soil-water tension does not include osmotic pressure and also called matric potential.
Solar constant	It is the amount of solar energy incident on a unit area at right angle to the sun's rays at the earth's mean distance per unit time in the absence of the atmosphere. His taken as cal/cm/min.
Solar energy	The energy transmitted from the sun in the form of electromagnetic radiation. Although the earth receives about one-half of 1 billionth of the total solar energy output, this amounts to about 420 trillion kilowatt-hours annually.
Solar farm	A suggested power utility, based in a desert and covering a considerable area, where large amounts of electrical energy would be generated from solar energy by capturing it by photovoltaic devices.

Term	Terminology
Solar radiation	It is radiant energy from the sun measured as a total amount expressed in call cm/hr.
Solar spectrum	It is the part of electromagnetic spectrum occupied by the wave lengths of solar radiation.
Solarization	Extremely high light intensities which have inhibitory effect on photosynthesis.
Sole crop	A crop grown in pure stand at optimum population (spacing).
Solubility	A measure of the amount of substance that will dissolve in a given amount of another substance.
Solubility of a fertilizer nutrient	The quantity of a given nutrient which will be extracted in a specific medium under specified conditions, expressed as a percentage by mass of the fertilizer.
Solubility product	For sparingly soluble salts (*i.e.,* those of which the solubility is less than 0.01 mol $= dm^3$). It is an experimental fact that the product of the total molecular concentrations of the ions is a constant at constant temperature. The product is termed the solubility product.
Soluble sodium percentage (SSP)	The proportion of sodium ion in solution in relatio relation to the total cation concentration.
Solum	It is the weathered portion of the soil (A and B horizons) mass, including the full depth to which illuviation and elluviation are in evidence.
Solute	The substances being dissolved in a solvent.
Solution	A homogeneous mixture formed by dissolving one or more substances (solid, liquid or gaseous) in another substance (solvent).
Solution Concentrate	A liquid formulation, also known as soluble concentrate, that consists of a herbicide dissolved in a solvent system, which forms a solution when added to water or a carrier.
Solution fertilizer	Liquid fertilizer free of solid particles.

Term	Terminology
Solvent	The substance in which the solute is dissolved. Sorghum injury: Poisoning of livestock resulting from feeding of young sorghum plants containing hydrocyanic acid (HCN). HCN concentration is more at seedling stage.
Sorption	It refers to surface-induced removal of a chemical substance from solution by adsorption, absorption or precipitation.
Sorting (seed)	It involves inspection, followed by separation according to variety, size, colour, degree of ripeness, maturity, defects, etc. It is actually a separating operation supplementary to grading.
Source	The term used to describe the organ of plant supplying the photosynthate, for example leaves.
Sowing	The process of placing seeds in seed bed.
Spacing	The distance between crop rows (inter-row spacing) and between plants within the row (intra-row spacing).
Spatial arrangement	It is defined as the pattern of distribution of plants over the ground, which determines the shape of the area available to the individual plant.
	The physical or spatial organisation of component crops in a multiple cropping system.
Specialized cropping	It is a cropping plan in which a single crop contributes 50% or more of the total production or monetary receipt (comparable equivalent) in 1 year.
Specific activity (nuclear)	The activity or decay rate of a radioisotope per unit mass of the sample (*e.g.,* microcuries/milli mole).
Specific crop-intensity index	Specific crop-intensity index determines the amount of area-time devoted to each crop or group of crops compared to total area-time available to the farmer for crop production during the time period under study.
Specific heat	The amount of heat needed to raise the temperature of 1 gram of the substance through 1°C.
Specific humidity	The ratio of the mass of water vapour in a volume of moist air to the total mass of the volume of moist air.

Term	Terminology
Specific ion toxicity	Any adverse effect from a salt constituent in the substrate on plant growth that is not caused by the osmotic properties of the substrate.
Specific leaf area	The leaf area per unit leaf weight.
Specific surface	The surface area per unit weight of soil, commonly expressed as m^2/g soil.
Spectrophotometer	Photometer adopted to measure the spectral distribution of the luminous flux in a beam of light.
Spikelet	It is the unit of the panicle.
Spill ways	A conduit in or around a dam for the escape of excess water.
Spillman's equation	According to Spillman plant growth increases with increasing application of limiting element but increase in growth is not in direct proportion to the amount of growth factor added. The increase in growth with each successive addition of the element in question is progressively smaller.
Splash erosion	The splattering of small soil particles caused by the impact of rain drops on wet soils. The loosened particles mayor may not subsequently be removed by surface run-off.
Split-plot design	A layout in which one set of treatments is assigned to large plots called main plots into which a replicate is divided and another set of treatments to subdivisions of the main plots called subplots is termed a split-plot design.
Spodosols	Soils with subsurface alluvial accumulations of organic matter and compounds of aluminium and usually iron. These soils are formed in acid mainly coarse-textured materials in humid and mostly cool or temperate climates.
Spoil bank	A pile of soil, subsoil, rock or other material excavated from a drainage ditch, pond of other cut.
Spot treatment	Application of spray to localized or restricted areas, as differentiated from overall broadcast, or complete coverage.
	An application of spray to a localized or restricted area.

Term	Terminology
Spray drift	The movement of airborne sprays particles from the spray nozzle beyond the intended contact area.
	The movement of airborne liquid sprays particles outside the intended area of application.
Sprayer	A machine used to apply fluids in the form of droplets of effective size and distribute them uniformly over the plants.
Spreading Agent	A substance used to improve wetting, spreading or possibly the adhesive properties of an herbicide spray solution.
Sprigging	Vegetative propagation by planting stolons or rhizomes (springs) in furrows or holes in the soil.
Sprinkler irrigation	Irrigation by means of sprinkler spaced at intervals on a pipe and is practiced under undulating topography, where water is scarce and cash crops are grown.
Sprout	To put forth new growth from seeds or seed pieces, *e.g.,* potato sprouts.
Spudding	Removal of weeds by cutting off below the soil surface.
Spur terrace	A short terrace used to collect or divert run-off.
Squall	Atmospheric phenomenon characterized by a very large variation of wind speed. It is often accompanied by thunderstorm.
Square	An unopened flower bud of cotton with its subtending bracts.
Square root transformation	Counts of events having a small probability of occurrence after follow poisons distribution, the data have to be made to follow normal distribution in order to apply to analysis of variance technique. The square root transformation is applicable under such circumstances, for the numbers between 0 to 10, the transformation $X \times 1/2$ is more nearly correct than X.
Stable isotope	An isotope which is not radioactive, *e.g.,* ^{15}N.
Stale seed-bed technique	Allowing weeds to germinate by rain or wetting and tilling to kill them before sowing seeds.

Term	Terminology
Standard deviation	A measure of dispersion of variable around the general mean and is a calculated by summing up the squares of the deviation of each observation from the mean and dividing by the number of observations and then extracting the square root.
Standard score	A score which represents the deviation of a specific score from the mean expressed in standard deviation units.
Starch	Consists of D-glucose units linked in a linear fraction by 1-4 linkages (amylose) and branched chain fraction by 1-6 linkages (amylopectin).
Starch equivalent	The number of kilograms of starch that are necessary to produce the same quantity of fat as 100 kilograms of the feed. Thus it expresses the fat-producing value of the feed in terms of the weight of starch.
Starter solutions	The term refers to solutions of fertilizers generally consisting of N-$P_2 O_5$ K_2O in the ratio of 1:2:1 and 1:1:2. These solutions are applied to young vegetable plants at the time of transplanting to help the plants to establish.
State farming	Farms are managed by government officials. In India all state farms are governed by an independent body, i.e., State Farm Corporation of India. Various research activities can be facilitated under state farming.
Statistic	Estimate of a parameter made from a sample. Statistics is to sample what parameter is to population.
Stefan Boltzmann's constant	It is the universal constant of proportionality between the radiant emmitance of a black body and the fourth power of the body's absolute temperature.
Stefan-Boltzmann law	One of the radiation laws which states that the amount of energy radiated per unit time from a unit surface area of an ideal black body is proportional to the fourth power of the absolute temperature of the black body; $R = W T^4$.
Stem-elongation stage	It occurs during late part of the tillering stage. In short-duration varieties it overlaps panicle-initiation stage.

Term	Terminology
Steppe	Vast semi-arid grass covered plain, usually lightly wooded.
Stevenson screen	It is a meteorological screen named after its inventor. It is meant for enclosing meteorological instruments as a protection from radiation and at the same time ensuring sufficient ventilation.
Sticky point (soil)	The moisture content expressed as a percentage of the oven-dry soil at which kneaded moist soil just ceases to adhere to external objects. It is a condition of consistency at which the soil barely fails to stick to foreign objects.
Stocking rate	Number of animal units per unit of land area at a specific time.
Stolon	Runners or stems that develop roots and shoots at the tip or nodes.
Stolon	The above-ground runner or stem that develops roots and shoots at the tip or nodes (*e.g., Cynodon dactylon*, bermudagrass).
Stoma (plural stomata)	An opening in the epidermal layer of plants for the passage of gases and water vapour.
Stomatal index	S. I. = No. of stomata/unit area ÷ No. of stomata/unit area + No. of epidermal cells/unit area × 100.
Storage (structure)	Storage structure that can keep the moisture content and relative humidity of the grains stable throughout the storage period.
Storage capacity (soil)	The amount of water that can be stored in the soil for future use by plants and evaporation.
Storage phase (irrigation)	Portion of the total irrigation time between the end of advance and inflow shut-off.
Stored soil water (Wb)	Depth of the water stored in the root zone from earlier rains, snow or irrigation applications which partly or fully meets crops water requirements in following periods; mm.
Stover	Matured, cured stalks of such crops as maize, sorghum or millets from which grains have been removed; a type of roughage.

Term	Terminology
Straight fertilizer	Qualification generally given to a nitrogen, phosphorus or potassium fertilizer having a declarable content of 1 primary nutrient only.
Straw	The dried remnants of crop plants which the seed has been removed, also called stover.
Stream lines	Path through which water flows in soil.
Stress	A potentially injurious force or pressure acting on the plant which may lead to injury or death.
Strip cropping	Growing soil-conserving and soil-depleting crops in alternate strips running perpendicular to the slope of the land or to the direction of prevailing winds for the purpose of reducing erosion.
Strip grazing	Confining animals to an area of forage to be consumed in a short period of time usually a day. It is also called ration grazing. In this system, a fixed or varying number of animals are given access to only part of a paddock by a movable fence; in addition a movable back fence may be used to prevent access to strips already grazed.
Strip intercropping	Growing 2 or more crops simultaneously in different strips wide enough to permit independent cultivation but narrow enough for the crop to interact agronomically.
Strip plot design	Two different sets of treatments can be tried in large plot with one set of plots superimposed over the other set at right angles. A block may be divided into strips in one direction to be allotted to 1 set of treatments and into another set of strips to be allotted to the second set of treatments, this design is used when two sets of treatments needs larger plots.
Strip tillage	A tillage system in which only isolated bands of soil are tilled.
Stubble	The basal portion of the stems of plants left standing after cutting or harvesting the crop.
Stubble mulching	Stirring the soil with implements that leave considerable part of the vegetative material or crop residues or vegetative litter on the surface as a protection against erosion and for conserving moisture by favouring infiltration and reducing evaporation.

Term	Terminology
Sub-irrigation	Irrigation in which measures are taken to control drainage discharge and to elevate the water-table locally to or nearly to the root zone by applying water in open ditches or tile drains.
Suberization	The formation of a complex wax like substance in the cell walls of wounded or cut plant tissue to reduce water loss and protect the area.
Sub-humid	A climate or region where the moisture content is below that under humid conditions but normally still adequate for the production of many crops without irrigation.
Submerged flow	The flow of water in which there is a back pressure as a result of the discharge is lower than the free flow.
Submergence	Partially or completely under water.
Subplot	A subdivision of an experimental plot or sample plot.
Subsistence crop	The crop which is grown under certain problematic conditions when no other crop can be grown, as floating-rice in flood prone areas.
Subsistence farming	It is farming enterprise which provides food and commodities just sufficient for the farming family, and there are no surpluses to sell.
Subsoil	That part of the solum below plough depth or below 'A' horizon.
Subsoil placement	Placement of fertilizers in the subsoil with the help of heavy power machinery usually followed in humid and sub-humid regions.
Subsoiling	Breaking the compact subsoils without inverting them with a special narrow cultivator, showel or chisel at depth usually from 30-60 cm from the surface.
Substituted urea	A compound wherein part of the hydrogen of the urea is replaced by other functional groups like methyl, phenyl etc.
Substitution cropping	It is substitution of an existing inefficient crop with an identified efficient crop in a zone or region.

Term	Terminology
Subsurface irrigation	Irrigation of crops by applying water below surface of the ground through pipes.
Succession	An orderly sequence of different plant communities in an area. One community replaces another until the climax in vegetation is reached.
Succulage	The root crops such as turnip, carrot and fodder beets which provide succulent livestock feed are called succulage. These crops are rich especially in carbohydrate and make good source for mixing with poor-quality roughages. Short-duration turnip and carrot are being extensively grown as 'catch crops' on large dairy farms.
Sucker	A laterally growing subterranean offshoot from the base of the main stem of a plant by which vegetative reproduction is affected.
Sugar crops	The crops grown for the purpose of producing sugar from them, *e.g.,* sugarbeet, sugarcane.
Sugarcane crusher	A machine used for extracting juice from sugarcane.
Sugarcane set	Small pieces of canes used for planting, usually taken from the top portion of the cane plant.
Sugarcane trash	Dried leaves of sugarcane.
Sulphur deficiency	Symptoms are chlorosis of younger leaves followed by yellowing of older leaves, stunted growth and reduced tillering. Sulphur deficiency occurs in soil low in organic matter in humid regions. Flooding aggravates sulphur deficiency by converting soluble sulphates to insoluble sulphides.
Sulphur-coated urea (SCU)	It has been developed as a slow release nitrogen fertilizer and is basically urea with a coating of elemental sulphur, including a binding agent, a sealant and a microbicide. The N content of SCU ranges from 10 to 37% depending on the thickness of the sulphur coating.
Sum of squares	Deviation from the mean squared and summed.
Sun-curing	The process of drying in sun, *e.g.,* sun-curing of tobacco.

Term	Terminology
Sunshine hours (n)	Number of hours of bright sunshine per day, also sometimes defined as the duration of traces of bums made on a chart by Campbell Stokes recorder; hours.
Sunshine recorder (heliograph)	Device which records the time interval during which solar radiation reaches the earth's surface with sufficient intensity.
Super cooling	Cooling below the freezing point without solidification.
Superimposed trial	A small set of experimental treatments superimposed on farmers' production plot or cropping pattern trial fields at a research site to evaluate the performance of alternative component technologies. These trials are usually managed by the researchers.
Superphosphate (single ordinary superphosphate)	A commercial product obtained by treating phosphate rock with sulphuric acid and containing about 16% of P_2O_5 mainly water-soluble, along with calcium sulphate and other products of reaction (contains 16% S).
Supplemental irrigation	The watering of crops in regions where normal rainfall ordinarily supplies most of the moisture. It is provided when crop is facing moisture stress due to dry spells.
Supplemental pasture	Additional pasture for use in adverse weather conditions, usually annual forage crops for dry periods or winter time.
Supplementary crops	Such crops are neither competitive nor complementary.
Supplementary enterprise	Those enterprises which do not compete in the use of resources and rather make use of the resources when they are not being utilized by other enterprises.
Support crops	Certain fast-growing crops which work as supported to vine crops, *e.g.*, castor.
Surface irrigation	An irrigation method wherein water is applied directly on the soil surface from a channel located on the upper side of the field: The water spreads over the fields by gravity flow.
Surface soil	The upper 12 to 20 cm of the soil, or in arable soils, the depth commonly stirred by the plough.

Term	Terminology
Surface Tension	A physical property of liquids, due to molecular forces, that causes them to form drops, rather than spread as a film.
Surface tillage	To till the surface without turning the tilled layers.
Surfactant	A material which in pesticide formulation imparts emulsifiability, spreading, wetting, dispersability or other surface-modifying properties.
Susceptibility	The sensitivity or degree to which a plant is injured by a herbicide treatment.
Suspension	A liquid or gas in which very fine solid particles are dispersed, but not dissolved.
Sustainable agriculture	Sustainable agriculture should involve the successful management of resources for agriculture to satisfy changing human needs while maintaining or enhancing the quality of the environment and conserving natural resources.
Sustained use	Continuing use of land without severe or permanent deterioration in the resources of the land.
Swamp	A completely or partially vegetated area subject to permanent inundation in stagnant or slowly flowing water.
Sward	The grassy surface of a pasture.
Sweath	A strip of cut herbage lying on the stubble.
Sweep	A double bladed V-shaped knife on a cultivating implement.
Symbiosis	An intimate physiological association of 2 or more species resulting in mutual benefit, *e.g., Rhizobium* bacteria on the roots of legumes.
Symbiotic nitrogen fixation	Fixation of atmospheric nitrogen by leguminous plants with the help of root nodule bacteria *(Rhizobium* spp.).
Symplast	A continuous system of protoplast comprising protoplasts of adjoining cells connected each other by means of plasmodesmata.
Synergism	Co-operative action of different chemicals such that the total effect is greater than the sum of the independent effects.

Term	Terminology
	Action of, two or more substances, in a mixture, with a total effect greater than either the sum of the two effects achieved independently or the predicted effect based on the response to each factor applied separately.
Synthetic cultivar	A cultivar developed by combining selected cross-pollinated plants or lines.
System synthesis	Selecting functional forms to quantitatively represent the cause-effect relationships and interdependencies among system variables in order to simulate, synthesize and forecast more nearly complete production systems. This, approach is, however, much more complex and provided many opportunities to make precise measurements on less relevant parameters that are potentially necessary to system analysis and evaluation.
Systemic Herbicide	A compound which is translocated readily within the plant and has an effect throughout the entire plant system; synonymous with translocated herbicide.
Systemic or translocated herbicide	Herbicide capable of moving within the plant to exert effects throughout the entire plant system irrespective of its place of entry, *e.g.,* 2, 4-D and' atrazine.
Systems approach	Studying a system as an entity made up of all its components and their interrelationships, together with relationships between the system and its environment.
T-distribution	Theoretical symmetrical sampling distribution with mean of zero and standard deviation which becomes smaller as degrees of freedom increase.
Tagging (Labeling)	Adding of a radioactive isotope to the reacting elements or compounds in a biochemical process.
Tal lands	Beyond the natural levees there are bowl-shaped depressions geologically known as back water which though in-undated are not subjected to erosion.

Term	*Terminology*
Tame pasture	A pasture covered with cultivated plants and used for grazing.
Tank silt	It consists of large proportion of finer soil particles of silt and clay and organic matter carried by run-off water from the surrounding soil to the tanks during heavy rains.
Tannin	A polyphenol compound found in tea leaves is the second most valuable constituent of tea, after caffeine.
Tap root	A root originating from the radical in dicot plants, having a prominent central portion growing vertically downward and giving off small lateral roots in succession.
Tarai region	Low elevation plains near foot-hills of Uttar Pradesh.
Target area	Large priority development area in a country for which improved cropping systems technologies are developed through on-farm cropping systems research conducted at sites in that area.
Tassel (maize)	The flower cluster at the tip of a corn plant comprised pollen-bearing flowers, *i.e.,* the staminate inflorescence of maize.
Taungya system	A type of agro-forestry system in which trees are planted with food crops, which serve as a nursery for the young tree seedlings, the aim is to produce a forestry or tree crop plantation, makes the early years of plantation establishment productive and reduce labour cost of establishment, *e.g.,* growing of cassava (Tapioca) for initial 2-3 years in *eucalyptus* plantation.
Taxonomy	The science dealing with the description, naming and classification of plants.
Tedder	An implement for stirring hay in the swath or windrow.
Temperate zone	Either of the 2 latitudinal zones on the earth's surface which lie between 23° 27' and 66° 27' N and S.
Temperature	Temperature is a relative term which indicates capacity to transfer heat by conduction.

Term	Terminology
Temperature inversion	An increase of air temperature with height so that warmer air overlies colder air.
Tempering (seed processing)	Bringing a product to a desired moisture content or temperature for processing. The term also applies to equalization of moisture or temperature throughout kernel or grain.
Temporary pasture	A pasture grazed during a short period only, usually not more than 1 crop season.
Temporary wilting	Wilting condition of the plant due to loss of water which would recover from wilted condition during the night without addition of any water.
Tensiometer	A device for measuring the negative pressure (or tension) of Water in soil in situ, consisting a porous permeable ceramic cup connected through a tube to a manometer or vacuum gauge. It is used in scheduling of irrigation.
Terrace	An embankment or ridge of earth constructed across a slope to control run-off and minimize soil erosion: Bench terraces and ridge terraces are the 2 general types.
Terrace interval	Distance measured either vertically or horizontally between corresponding points on 2 adjacent terraces.
Terrestrial radiation	The long-wave radiation emitted by the earth's surface.
Terrestrial weeds	Weeds which normally start and complete their life-cycle on land.
Test of significance	Statistical test designed to distinguish differences due to sampling error from differences' due to discrepancy between observation and hypothesis.
Test sample	A representative part of the final sample prepared by an appropriate method for a particular test.
Test weight	It denotes the weight of 1,000 grains (normally) of field crops. In some crops 100 seed weight in also considered.
Testa	Hard outer seed-coat developed from the integument(s) of an ovule.
Tester animals	Animals used in grazing experiments to measure animal performance or pasture quality.

Term	***Terminology***
Texture	The relative proportion of the various size groups of individual soil grains or separates.
Thermal conductivity	The quantity of heat flowing per unit time through a 1 cm^2 cross section of the soil in response to a temperature gradient of 1°C cm^{-1} of depth.
Thermal diffusivity	It is the change in °C that occurs in one second when temperature gradient changes 1° per sec.
Thermal induction	The change in growth and development of plants brought about by a given temperature exposure, *e.g.*, vernalisation.
Thermal neutrons (nuclear)	Neutrons having energies corresponding to room temperature, that is approximately 0.025 eV which is the kinetic energy of the molecule at about 300°K.
Thermo isopleths	Lines of equal soil temperatures.
Thermocouple	A device that uses the voltage developed by the junction of two dissimilar metals to measure temperature differences.
Thermograph	Apparatus used to give a continuous record as a graphical chart of temperature with time.
Thermo-hygrograph	Instrument resulting from the combination of a hygrograph and thermograph. It furnishes on the same chart simultaneously a continuous recording of atmospheric temperature and humidity
Thermoperiodicity	Effect of temperature difference between light and dark periods upon plants.
Thermoperiodism	The term thermoperiodicity or thermoperiodism designates the effects of alternations of temperature between the day and night periods upon the reactions of plants.
Thinning	Removing of extra plants from thickly populated crop stand with an idea of maintaining optimum plant population.
Three-dimensional diagrams	A form of graphical representation of numerical data through diagrams which are in the form of Cubes. The volume of the cubes is proportional to the magnitude of the variate.

Term	Terminology
Threshing	The process of detaching the grain from the earheads or from the straw.
Throw (soil)	The movement of soil in any direction as a result of kinetic energy imparted to the soil by tillage tool.
Tidal swamps	Low elevation coastal area affected by tidal flooding.
Tile drainage	Pipe made of burned clay, concrete or ceramic material in short lengths, usually laid with open joints at suitable spacing and depths in the soil or subsoil to collect and carry excess water from the soil.
Till	To plough or cultivate soil for seeding.
Tillability	The degree of ease with which a soil may be manipulated for a specific purpose.
Tillage	The use of implements for mechanical manipulation to prepare seed-beds conducive for field crop production.
Tillage action	The action of a tillage tool in executing a specific form of soil manipulation, such as soil cutting, pulverizing or inversion.
Tillage depth	Vertical distance from the initial soil surface to a specific point of penetration of the tool.
Tiller	Side-shoot growing from the base of the stem (at ground level) of a cereal or grass plant.
Tillering stage	The appearance of the first tiller from the axillary bud in one of the lowermost nodes.
Tilth	The physical condition of the soil with respect to its fitness for the planting or growth of a crop.
Tissue analysis	Quantitative estimation of the nutrient content of plants carried out by suitable set procedure such as drying the plant part and analysing it by chemical means in the laboratory.
Tissue culture	The growth of detached pieces of tissue in nutritive fluids under asceptic conditions.
Tissue testing	Sampling a small portion from a growing plant for nutrient or chemical estimation. Petioles and tender stem portion are usually used in tissue testing for rapid estimation.

Term	Terminology
Titer test	It is when molten fatty acids are cooled and begin to solidify, the latent heat of fusion is liberated and a sudden rise in temperature can be observed. The highest temperature recorded is known as the titer point.
Tocopherols	These are methylated derivatives of tocol, and are widely distributed in vegetable lipids. They have vitamin E activity and can protect unsaturated fats against oxidation.
Tolerance	Ability to withstand herbicide treatment without marked deviation from normal growth or function. The concentration of herbicide reside that will be allowed in or on agricultural products.
	In the context of herbicide tolerance, it is defined as the natural or normal variability of response to herbicides that exists within a plant species and can easily and quickly evolve.
Tolerance (pest/chemical)	The ability of a plant to sustain effect of a pest or chemical without getting seriously damaged.
Tolerance level (pesticide)	The safe limit of pesticide or chemical allowed by law to be in or on the plant or animal product sola for human consumption.
Top feed	Leaves of trees or shrubs which are used for feeding livestock are called top feeds and the species are often termed as top species. The loppings from such species serve as supplementary source of feed in arid and semi-arid areas, especially during crucial lean period when grosses are either grazed away or become dry with poor digestibility.
Top soil	The surface soil, usually up to the plough depth (15-20 cm from the surface).
Top-dressing	The application of manure or commercial fertilizer to a crop that is already established.
Topoclimatology	Study of local climate as affected by slope and other surface characteristics.
Topography	The configuration of surface including its relief and the position, of its natural and man-made features.

Term	Terminology
Topping (tobacco)	The removal of the terminal bud with or without some of the small top leaves just before or after the emergence of a flower bud.
Total digestible nutrients (TDN)	Sum total of all digestible organic nutrients, *i.e.,* proteins, nitrogen free extract fiber and fat.
Total nomadism	In this system, the animal owners do not have a permanent place of residence. They do not practice regular cultivation, and their families move with their herds.
Total radiation	Sum of solar and terrestrial radiation.
Toxicity	Harmful effects of a substance on plants due to its excess absorption/application above the tolerance limit.
Toxicology	The study of the principles or mechanism of toxicity.
Toxin	A poisonous secretion of a plant or an organism.
Trace (nuclear)	The test object, element or compound that the investigator is endeavouring to trace.
Trace elements	An old term used for 'micro-nutrients or minor elements' which are taken up in small quantities by a crop plant?
Tracer (nuclear)	An isotopic tracer is an isotope used to tag or follow a chemical reaction or process such that its location and concentration can later be determined.
Tracer element	The isotope of an element which may be incorporated into a sample to trace the course. Of the element, alone or in combination, though chemical, biological or physical processes.
Trade name	A trademark applied to herbicide formulation by its manufacturer.
Trans-amination	An important reaction in amino add synthesis which involves the transfer of an amino group of an amino acid to the carboxyl group of a keto acid giving rise to a corresponding amino acid.
Transcription	Formation of messenger RNA from a DNA template which carries the genetic information to the ribosome for a protein synthesis.

Term	Terminology
Transformation of data	The assumptions underlying the analysis of variance and the usual tests of significance stipulate the normal distribution of the data. Whenever the data do not follow the normal distribution the variate X_1 transformed to some new variate say $f(X_1)$ in such a way that $f(X_2)$ is normally and independently distributed with mean μ and variance W^2.
Translation	The m-RNA through its message to ribosomes causes the synthesis (translation) of polypeptide chain and this process is known as translation.
Translocated herbicide	An herbicide that is moved within the plant. Trenslocated herbicides may be either phloem mobile or xylem mobile, but the term is frequently used in a more restricted sense to herbicides that are applied to foliage and moved downward through the phloem to underground parts.
Translocation	Transfer of food or other materials such as herbicides etc. from one plant part to another.

Transfer of photosynthates and other materials such as herbicides from one plant part to another. |
Transmission and reflection	Radiant energy that is not absorbed is either transmitted or reflected. Transmission or reflection is usually expressed as decimal fraction or as percentage.
Transmittance ('I') or per cent transmission	This is the fraction of light transmitted by a substance. It is expressed as a decimal fraction, T = 1/10 or a percentage (1/10 x 100), where 10, irradiance of incident radiant energy and I, irradiance of transmitted radiant energy.
Transmutation (nuclear)	A process in which one target nucleus is transformed into a compound nucleus.
Transpiration	The process of water vapour release to the atmosphere from the aerial organs of the plant.
Transpiration coefficient	The quantity of water (g) necessary for a plant to produce 1 kg of dry matter.
Transpiration pull	A capillary pull arising out of a fall in pressure as a result of transpiration at the top of the plant.

Term	Terminology
Transpiration ratio	The quantity of water transpired to produce a unit amount of dry matter.
Transplanter	A machine which sets seedlings into the grown in rows. The spacing between the seedlings is adjustable.
Transplanting	The process of planting seedlings in prepared seed-beds.
Transported soil	Soil formed by the consequent or subsequent weathering of material transported and deposited by some agency such as water, air etc.
Transverse drainage	A method of drainage where the drains are placed in a direction more or less at right angles to the direction of the steepest slope of the land to be drained.
Trap crops	Grown to trap insect-pests and soil-borne harmful biotic agents such as parasitic weeds.
Trartsplant shock	The rate of transpiration is more than absorption in the just-transplanted seedlings and such seedlings undergo a kind of shock called transplant shock.
Trashing	Stripping of lower leaves of sugarcane crop is known as trashing.
Tratogenic	Capable of producing birth defects.
Treatment	A single state of some factor being varied in an experiment, such as a rate of herbicide, method of applying a herbicide, sources of fertilizer etc.
Tree cropping	It is based on the concept of using tree crops as the basis of an agricultural system. It entails replacement of grain crops with tree crops, consisting of an upper tree storey of one or more types of forage trees, a ground level permanent pasture and one or more type a of domestic animals that feed on the pasture or on tree products.
Trial (experiment)	A group of plots to which treatments have been applied, arranged in a manner dictated by the chosen experimental design.

Term	Terminology
Trickle irrigation	A method of applying water directly near the root zone of the plants through a number of low flow-rate outlets generally placed at short intervals along small tubing. Sometimes it is referred to as drip irrigation followed for high value crops in arid regions.
Trigger factor	A factor, the change of which sets off a chain reaction in an ecosystem.
Triple cropping	Growing three crops a year in sequence.
Triple superphosphates	A commercial product obtained by treating phosphate rock with phosphoric acid and containing about 46% P_2O_5, mainly water-soluble.
Triticale	It is a man made genus created to produce a new cereal with a combination of the characteristics of wheat and rye. It is an allopolyploid between bread wheat *(Triticum aestivum)* and rye *(Secale cereale)*. It has large spike accommodating more grains than wheat and is highly tolerant to moisture stress.
Trophic level	The functional level occupied by an organism in any food chain.
Tropic movements	Movements in plants which occur under the influence of environmental factors that act with a greater intensity from one direction than from another, *e.g.,* phototropism.
Tropical region	Comprises the area between the Tropic of Cancer (23°27' N latitude) and the Tropic of Capricorn (23°27' S latitude).
Tropism	A growth curvature induced by a stimulus, such as light or gravity.
Truck crops	Crops which yield in tonnes and grown for distant market requiring heavy transport.
True density	The mass per unit volume of the particles of a material.
True digestibility	Actual digestibility or availability of a feed, forage or nutrient as represented by the balance between intake and faecal loss of the same ingested material. Thus the true digestibility of a feed by animals is the apparent digestibility.

Term	Terminology
Truncated soil	Soil which has lost material from the upper part of the profile by the processes of erosion.
Tuber	Swollen underground stem or root which serves as a food store, and is an organ of vegetative reproduction. Stem tubers may be recognized by the buds which are formed in axils of rudimentary leaves, *e.g.,* potato. Root tubers have lateral roots emerging from the sides *e.g.,* sweet potato.
Tuber crop	Crops possessing enlarged underground reproductive stems, roots, *e.g.,* potato, cocoyam, sweet potato etc.
Turbulence	A state of fluid flow in which instantaneous velocities exhibit irregular and apparently random fluctuations.
Turf	The upper stratum of soil filled with the roots and stems of 10w-growing plants, especially grasses.
Turgid	Describing a cell or tissue which is firm and plump because of the internal pressure resulting from the osmotic uptake of water.
Turgor	The swollen conditions of the cell resulting from the hydrostatic pressure exerted by the protoplast on the cell wall.
Turgor movement	Movements of plant organs caused by reversible change in cell volume due to change in turgor.
Turgor pressure	The pressure exerted on the cell wall by the cell contents inside it when the cell contents are fully turgid.
Turn around time	A period between the harvesting of the preceding and planting of the succeeding crop in a specific field:
Turn-out valves	These are solid iron structures installed in cement concrete base in *pukka* channels which are used for opening and closing water supply for irrigation. It is a water control device.
Type (plant)	A group of varieties of similar characters or an admixture of seed of indistinguishable varieties but having some characteristics in common, like grain type sorghum, yellow maize etc.

Term	Terminology
Type II error (type error)	The probability of accepting a hypothesis when it is actually false.
Type terror (type error)	The rejection of hypothesis when it is actually true.
Ultisols	Soil that are low in bases and have subsurface horizons of alluvial clay accumulation. They are usually moist, but during the warm season of the year some are dry, part of the time.
Unit value of fertilizer ingredients	One % of N, P_2O_5 and O present in 1 tonne of a fertilizer is called 1 unit. The unit value of the fertilizer divided by the percentage content of that nutrient.
Universal soil loss equation (USLE)	An empirical relationship to describe the factors which affect soil loss by erosion.
Unsaponifiable matter	It includes all those constituents of fats which are not saponified by alcoholic caustic potash and which are soluble in petroleum ether or ethyl ether.
Unsaturated fatty acids	Fatty acids which contain one or more pairs of carbon atoms linked by double bond. These can be saturated by breaking the double bonds through hydrogenation: *e.g.,* oleic acid, linoleic acid, linolenic acid.
Upland	Cultivable land situated on high ground or the land which has perfect natural drainage.
Upland rice	Rice grown in high level land in rainfed, naturally well-drained soils, without surface water accumulation, normally without pheratic water supply and normally not bunded.
Upper and lower moisture plastic limits	The upper limit is that moisture content at which the soil will barely flow under an applied force. The lower plastic limit is that moisture content at which the soil can be rolled out into wire.
Urea [$CO(NH_2)2$]	A synthetic, non-protein organic compound, crystalline or made into granules or prills for fertilizer use and containing 46% nitrogen.
Urea formaldehyde	Slow-release nitrogen fertilizer produced by reaction between urea and formaldehyde.

Term	*Terminology*
Urease	A universally occurring enzyme which activate urea hydrolysis to produce CO_2 and NH_3.
Utilization	It is the extent to which grazing animals have chopped the current growth within their reach, expressed in percentage of height or weight of forage removed or in such general terms as excessive, close, moderate or light.
Value or Et-value	It is a measure of the amount of isotopically exchangeable phosphate of the soil and like L-value, the E-value measures labile P, made up of' Surface-p and Solution-p. It is the laboratory equivalent of the 'L' value.
Vapour	It describes a substance in the gaseous stage below its critical temperature.
Vapour drift	The movement of vapours from the area of application to other areas.
Vapour pressure	The pressure exerted against the walls of the container and the surface of the water by the water-vapour molecules.
	The property which causes a chemical compound to evaporate or vapourize.
Variable	A characteristic or phenomenon which may take on different values.
Variance	Measure of deviations of variates from their mean. The square of the standard deviation. The corresponding statistic is the mean square.
Variate	A single observation or measurement.
Variety	A subdivision or a group of plants within a species which is characterized by growth, plant fruit, seed or other characters by which it can be differentiated from other seeds of the same kind.
Vascular Tissue	A general term referring to either or both xylem and phloem.
Vegetation	Sum total of plants in a given area.

Term	Terminology
Vegetative cover	A soil covers of plants irrespective of species. Vegetative propagation: The production of a complete plant by asexual methods like grafting, layering, cutting, division, separation etc.
Vegetative stage	Growth stage of a crop plant prior to flowering.
Vermicompost	Compost produced from the digestion and discharge of refuse and other organic wastes by earthworms.
Vermiculture	It means artificial rearing or cultivation of earth worms.
Vernalization	Low temperature treatment of seeds and seedlings, which has the effect of artificially 'ageing' the plant and reducing the time between sowing and flowering.
Vertical mulching	Incorporation of vegetative mulch in a band in the soil for the purpose of harvesting rain water and conserving soil and water.
Vertical revolution in agriculture	Maximizing production per unit land area per unit of time using intensive cropping systems, high production inputs and improved management practices.
Vertisols	Clayey soils with high shrink-swell potential that has wide, deep cracks when dry. Most of these soils have distinct wet and dry periods throughout the years.
Viability	The capability of a plant structure (seed, cutting etc.) to show living properties like germination and growth.
Viability of seeds	It is represented by germination percentage which expresses the number of seedlings which can be produced by a given number of seeds.
Vigour	A condition of active good health and material robustness in seeds, which upon planting permits germination to proceed rapidly and to compete under a wide range of environmental conditions.
Virgin soil	A soil that has not been scantly disturbed from its natural environment.

Term	Terminology
Virus	A non-cellular entity which consists minimally of protein and nucleic acid (DNA or RNA) and which can replicate only after entry into specific type(s) of living cell.
Visible radiation	Radiation from the violet at about 380 nm to the far-red at about 770 nm.
Vitamins	Substances required in minute amount for the normal metabolism of plants and animals.
Void space	The space between the particles in a bulk of stored grain, usually expressed as per cent of total volume.
Volatile	A compound is volatile when it evaporates or vaporizes (changes from a liquid to a gas) at ordinary temperatures on exposure to the air.
Volatilization	A process where a condensed phase such as liquid or solid, is transformed .into vapour by elevation of temperature or reduction of extemal pressure; also described as a measure of the tendency to evaporate or vapourize (from a liquid to a gas) at ordinary temperatures on exposure to the air.
Volume weight of soil	It indicates the number of times heavier a given volume of dry soil is than that water that will occupy the same total soil volume.
Volunteer plants	Unwanted plants growing from seed that remains on the field from a previous crop.
Wage	The amount paid periodically to the labourer for the work done in the field.
Ware crops	Such crops are grown for temporary storing a intact in warehouse for future use or sale.
Warm-season crop	A crop species that grown during the warmer part of the year.
Waste-lime (By-product lime)	An industrial waste or byproduct containing calcium and magnesium in forms that will neutralize acids. It may be designated by prefixing the name of the industry or process by which it is produced; for example gas lime
Water balance	It is hydrological balance between rainfall plus irrigation to that of changes in soil moisture, evaporation, percolation and run-off.

Term	Terminology
Water compaction	Water furnished by destruction of pore space owing to compaction of sediments.
Water conservation	Physical control, protection, management, and use of water resources in such a way as to maintain growing crop and forest lands, vegetation cover, wildlife habit and maximum sustained benefits to people, agriculture, industry, commerce, and other segments of the national economy.
Water culture	An experimental means of determining the mineral nutrient requirement of a plant; the plant is grown with its roots dipping in solution of known composition.
Water deficit	Shortage of water or inadequate water supply. The amount of water needed to return the soil-moisture status to the field capacity (FC).
Water furrow	A shallow trench or depression between 2 raised soil beds in which irrigation water or surface drainage can flow.
Water gates	A water-control device, made of wooden planks or metal sheet and used for opening and closing water supply. It is placed in the head ditch near farm or in field ditches.
Water harvesting	Conservation of rain water under un irrigated condition, by collecting run-off of precipitation in order to supplement soil moisture in an adjacent area.
Water potential	The capability of soil water to do work compared with free-water. The water potential at the surface of free-water is taken as zero.
Water quality	It refers to the characteristics of water supply that will influence its suitability for a specific use, like for irrigation of crops; usually the emphasis is placed on the chemical and physical characteristics of the water.
Water regimes	Regime means level; different levels of water (standing or flooding) depth applied in irrigation experiments. Such as 2.5, 5, 7.5, 10 cm etc. standing water to saturation or field capacity (FC) to rice crop etc.

Term	Terminology
Water requirement of crops	It is that quantity of water regardless of its source; required by a crop in a given period of time for its maturity, and it includes losses due to ET plus the unavoidable losses during application of irrigation water and water required for special operations such as land preparation, paddling, leaching etc. and is expressed in depth for given time.
Water saturation deficit	Parameter for measuring water deficit in plants.
Water shed-based fanning system	It involves the optimum utilization of the catchment precipitation through improved water, soil and crop management directly through infiltration of monsoon rainfall after run-off collection and storage or after deep percolation recovery from well for the improvement and stabilization of agriculture on the watershed.
Water yield	It is the total annual run-off volume.
Water-application efficiency	The percentage of water applied that can be accounted for as moisture increase in the crop-root zone of the soil.
Water-application rate	The rate in cm depth or ha-cm/hr that irrigation water is applied to fields.
Water-conveyance efficiency	It is the proportion of water delivered to the irrigated field from the total water diverted: from the source.
Water-distribution efficiency	The extent to which water is' uniformly distributed in an irrigation system.
Water-holding capacity	The weight of water held by a given quantity of absolutely dry soil when saturated.
Water-logged	A condition of land where the ground water soil water content reaches a level that is detrimental to plants and other useful soil microflora.
Watershed	It is the area above a given point on a stream that contributes water to the flow at that point. Catchment basin and drainage basin are synonymous with it.
Watershed measurement	The planned use of watershed lands in accordance with predetermined objectives, such as the control of erosion, stream flow, sedimentation and improvement of vegetative cover and other related resources.

Term	Terminology
Watershed treatment	Include all practices applied to the land that are effective in reducing flood runoff, controlling erosion, and increasing the amount of surface storage, rate of infiltration and water holding capacity of the soil.
Water-stable aggregate	A soil aggregate stable to the action of water such as falling drops or agitation as in wet-sieving analysis.
Water-table	The upper surface of ground water or that level below which the soil is saturated with water.
Water-use efficiency	The amount of dry matter that can be produced from a given quantity of water.
Waterway	A natural course for the flow of water.
Watt	It is a unit of measurement of power in the MKS system and equals 1 J/sec. One horse power is (hp) equivalent to approximately 746 watts.
Wave length of light	The distance between 2 corresponding points on any 2 waves, usually measured in Angstrom units
Weather	The state of the atmosphere with respect to wind, temperature, cloudiness, relative humidity, pressure etc. at a given time.
Weathering	The physical and chemical disintegration and decomposition of rocks and minerals during soil genesis.
Weed	A plant growing where it is not desired.
	Any plant that is objectionable or interferes with the activities and welfare of humans
Weed control	The process of limiting weed infestations, so that crops· can be grown profitably or other operations can be conducted efficiently.
Weed eradication	The complete elimination of all living plants, plant parts, and seeds of weed infestation from an area.
Weed index	It is an index expressing the reduction in yield due to the presence of weeds in comparison with weed-free situation.

Term	Terminology
Weed Management	Rational development of appropriate technology to minimize the impact of weeds, provide systematic management of weed problems, and optimize intended land use.
Weed Prevention	Stopping of weeds from invading and contaminating new area.
Weed spectrum	Refers to different kinds of weed species a particular herbicide will kill or inhibit.
Weed-control efficiency (W.C.E.)	It denotes the efficiency of the applied herbicide or a herbicidal treatment for comparison purpose.
Weeding	Removal of weeds (unwanted plants) from the field.
Weir	An obstruction placed in a stream, diverting the water through a prepared opening for measuring the rate of flow.
Wet land	Pertaining to soils flooded for at least several weed each year, or to crop grown in such soils.
Wet planting	A method of planting in which irrigations given before planting and is usually followed in sandy soils.
Wet season	A period during which precipitation in excess of water requirement and water accumulated in the soil and in reservoirs.
Wet year	If the rainfall exceeds twice the normal deviation at a particular place, that year is said to be a wet year.
Wet-bulb temperature	Temperature recorded on a thermometer whose bulb is surrounded by a wet muslin bag, thus lowering the temperature by loss of latent heat through evaporation; degree Celsius.
	It is one of the 2 thermometers of a psychrometer whose bulb remains wet with thin film of distilled water and indicates wet-bulb temperature.
Wet-bulb. depression	Difference between simultaneous readings of wet and dry-bulb thermometers; degree Celsius.
Wet-seeded rice	Sowing of soaked rice seed in standing water (7-15 cm depth) on well-prepared seed-bed.

Term	Terminology
Wettable powder	A finely divided dry formulation that can be readily suspended in water.
Wettable Powder	A finely divided dry formulation (powder) that will readily form a suspension in water.
Wettable powder (pesticide)	Wettable powders are dry free flowing powders containing a high concentration of (25% to 75%) of an active ingredient to which dispersing agents and wetting agents are added to facilitate the dispersion of wettable powder in extending medium such as water.
Wetted perimeter	The surface of a water conduit which is in contact with flowing water.
Wetting agent	A compound which when added to a spray solution causes it to contact plant surfaces more thoroughly.
	A compound which when added to a spray solution causes it to spread over and wet plant surfaces more thoroughly due to reduction in inter-facial tensions.
Whirling psychrometer	Psychrometer which is attached to a string or small chain which an observer can rotate in order to ensure good ventilation.
Whole-farm analysis	A methodology designed to search for optimal solution through incorporation of farmers' objectives, farming systems, and resources to arrive at improved cropping and livestock patterns and management practices for overall farming systems performance.
Whole-farm approach	An essential characteristic of farming systems research and development in which teams look at a whole farm to identify problems, opportunities, and interrelationships, to design and conduct experiments, and to evaluate results.
Wild flooding method	A method of applying water to land by overflowing field ditches from which it spreads over the land to be irrigated, controlled only by the slope of the land.
Wilt	Loss of freshness and drooping of foliage of a plant due to inadequate supply of moisture, excessive transpiration, or by a disease which interferes with the utilization of water by the plant.

Term	Terminology
Wilting point	It is soil-moisture condition at which the release of water to the plant roots is just barely too small to counterbalance the transpiration losses. On an oven-dry basis it is the moisture content of soil at which plants wilt and fail to recover their turgidity when placed in a dark humid atmosphere. It is also called wilting coefficient or permanent wilting point.
Wind	Wind refers to air in motion, is best expressed in km/hr. It is measured by anemometer.
Wind blast	Injury to leaves and twings by strong winds.
Wind break (shelter belts)	A strip of trees or shrubs or crop plants serving to reduce the force of wind and provide a protective shelter against wind.
Wind energy	It is also a form of solar energy since winds are caused by variations in the amount of heat that sun sends to different parts of the earth. Electricity is produced when a wind mill catch the wind and revolves, rotating a turbine which powers electric generator.
Wind erosion	The detachment, transportation, and deposition of soil by the action of wind. The removal and redeposition may be in more or less layers or as localized below out and dines.
Wind generators	Devices to extract energy from the wind to be used to generate electricity directly.
Wind mill	Machine with a motor that is moved slowly by the wind to produce mechanical power, used originally to mill grain and pump water.
Wind power	A power source derived from the use of windmills which converts the wind's energy into electricity. The power contained in a moving wind stream is proportional to the first power of both the air density and the area of the wind stream and proportional to the third power of wind velocity.
Wind speed	Speed of air movement at 2 m above ground surface in unobstructed surrounding. Mean in m/see over the period considered, or total wind run in km/day.

Term	*Terminology*
Wind strip cropping	The production of crops in long, relatively narrow strips placed crosswise of the direction of the prevailing winds irrespective of the contour of the land.
Wind-energy conversion (WEC)	A wide variety of machines have been designed to convert wind energy into useful power. The large scale WEC system currently being developed to utilize a horizontal design.
Windrow	A row of material formed by combining 2 or more swaths.
Windrower	A machine to cut crops and deliver them in a uniform manner in a row.
Wind-vane	Device used to indicate the directions from which the wind is blowing.
Winnower	A machine with 1 or 2 sieves and fan using air stream across falling grain to separate seed and chaff.
Winnowing	The process of separation of grains from the mixture of grain and chaff or straw or bhusa.
Winter annual	The plant that starts germination in the fall (after the summer), lives over the winter and completes its growth, including seed production, the following season.
Winter killing	Relatively high water transpiration rates in evergreens during a period when absorption of water can proceed only at a relatively slow rate may lead to a type of cold injury called winter killing.
Wood ash	A source of potash used as potassium manure containing about 2% P_2O_5 and > 5% K_2O.
Wood gas	Gas produced during production of charcoal by heating wood in the absence of air and usually used as a fuel at the production site.
Woody Plants	Plant that develop woody tissues
Work	Energy transferred from one body to another in such a way that a difference in temperature is not directly involved. Work is measured in terms of the amount of force multiplied by the distance it has travelled. The practical unit is the foot-pound.

Term	Terminology
Working day	The day considered for the purpose of work is normally 8 hours.
Xanthophyll	A yellow or orange cartooned pigment associated with chlorophyll in chloroplasts, also present in certain chromoplasts.
Xerophyte	A plant with structural and physiological features which permit it to grow in a dry habitat, or under water-stress conditions.
Xerophytism	The ability of plant to withstand dry conditions.
X-rays	The electromagnetic radiation in the region below 100 angstroms.
Xylem	The woody portion of the conducting tissue in the plant specialized for the conduction of water and minerals.
	The principal water-conductized non-living tissue in vascular plants characterized by the presence of tracheids; may also contain vessels, parenchyma cells, fibers and sclereids.
Xylophyte	A woody plant.
Yield components (yield attributes)	These are the components which finally make up or control yield of any crop, for, example in cereal grain weight/grain, grain number/ear, ear number/plant and plant number/unit area.
Yield maximization	Agronomic practices adapted to get the highest possible crop production per unit area per unit time without considering either the cost of production or net return.
Yield potential	Full production capability of a crop.
Yield stability	The maintenance of yield at desired level over a period of time. Its index is measured by inverse of coefficient of variation, regression between yield and environmental index and disaster level estimated by probability of system.
Yields of crop	The harvested produce obtained from a crop grown in a unit area of land, usually expressed as quintals or tonnes per hectare.

Term	Terminology
Zaid	A relatively short crop season (April-June) in between the two main seasons *Rabi* and *Kharif.*
Zero-plane displacement	An empirically determined constant introduced into the logarithmic velocity profile to extend its applicability to very rough surfaces or to take in to account the displacement of a profile above a dense crop.
Zero tillage (Ro-till)	The extreme in conservation tillage is no-tillage (zero tillage) wherein the new crop is planted in the residue of the previous crop without any prior soil tillage or seed-bed preparation and is usually possible when all weeds are controlled by the use of herbicides.
Zinc deficiency	Symptoms appear 2-4 weeks after sowing as blanching of the midrib of the emerging leaf, especially at base. Brown spots appear on the older leaves, which later on coaheses to give brown color. Tillering and growth are depressed. In severe deficiency the plant dies. Zinc deficiencies associated with calcareous, alkali, peat, and volcanic soils and soils that remain wet or waterlogged most of the year. Incidence is more severe where high rates of nitrogen and phosphorus are applied.
Zygote	A fertilized egg formed after the fusion of male and female gametes.

4

DIFFERENCES / COMPARISON

Difference between soil fertility and soil productivity

	Soil fertility	Soil productivity
1.	Considered as an index of nutrient avail ability to plants.	Usually, used to indicate the ability of the soil for crop yield.
2.	One of the factors of crop production (water, solar radiation etc.).	Interaction of crop production factors.
3.	Usually assessed in the lab.	Assessed in the field with reference to a particular climate.
4.	Soil potential to produce a crop.	Result of different factors influencing soil, management.
5.	Depends on physical, chemical and biological factors of soil.	Depends on, soil physical conditions and fertility, climate and weather etc.
6.	Function of available nutrients in the soil.	Function of soil fertility, soil and crop management and climate.
7.	Fertility of certain soils may be same in all the climates.	Differs in response to variation in climate and location.
8.	Soil fertility = f(nutrient status of soil)	Soil productivity = f(soil fertility + manage ment + climate).

Compare between existing research and FSR

	Existing Research	Farming Systems Research
1.	Existing one is completely component based research and there is no system.	Unlike component research it is system based research.
2.	Inappropriate technology of needs and circumstances of practicing farmer.	It focuses on generation of low cost, location specific relevant technology.
3.	Faulty top-bottom conventional research extension method.	It lays emphasis on farmers' participatory research and ensures smooth diffusion and adoption.
4.	Do not consider biophysical, socio-economic and other production constraints of farmers.	It enhances productivity provides full employment to farm family member & focuses mainly on benefit of resource poor farmers.
5.	Not-eco-friendly and sustainable technology	It is based on environment friendly and sus-tainable.

Difference between dry bulb and wet bulb

Dry bulb temperature	Wet bulb temperature
1. Mercury in glass thermometer called by dry bulb thermometer is used for measuring dry bulb temperature (air temperature).	1. Temperature on cool air (wet bulb temperature is measured with wet bulb thermometer. It is similar to dry bulb thermometer, but the bulb of the thermometer acts as evaporating surface.
2. The temperature range is from −35 to 55°C. Least count is 0.5°C. However, observation can be recorded up to 0.1°C. This temperature is used for calculating humidity, vapour pressure and dew point.	2. The bulb of the thermometer is continuously kept moist by muslin cloth covering the bulb. Four strands of cotton thread placed in a small container with distilled water keeps the muslin cloth covered bulb continuously wet.

Difference between weather and climate

Weather	Climate
1. Pertains to the day-to-day State of the atmosphere at a particular place.	1. Pertains to the atmosphere over a given region.
2. Refers to specific instant of time and place.	2. Refers to a large region and for a long period of time.
3. It is always changing and differs from time to time.	3. It is more or less stable and differs from region to region.
4. Timing of land preparation, planting date, plant protection.	4. Choice of farming systems and crops.
5. Choice of alternate cultivars.	5. Choice of ideal cultivars.
6. Choice of equipment for day-to-day needs.	6. Choice of farm equipment.
7. Contingent drought management.	7. Drought proneness of the region.

Difference between cyclone and anticyclone

Cyclone	Anticyclone
1. Low pressure in center.	1. High pressure in center.
2. Circulation is anticlockwise in northern hemisphere.	2. Circulation is clockwise in northern hemisphere.
3. Relatively more destructive.	3. Relatively less destructive.
4. Change in weather is speedy and for short duration.	4. Change in weather is low and mild and remain for long duration
5. Cirrus clouds formmation.	5. Cirrus clouds absent.
6. More cloudiness and rainfall.	6. Less cloudiness.
7. Inward spiraling of air.	7. Air moves outward from center.

Difference between straight fertilizers and mixed fertilizers

Straight fertilizers	*Mixed fertilizers*
1. It gives only one nutrient to the plant.	1. It gives several nutrients depend upon mixture of fertilizers.
2. Its cost may be higher than the mixed fertilizers.	2. Cost may be reduced due to mixing other nutrients together.
3. It can be mixed with other fertilizers but depend upon their mixing properties.	3. Mixing of various fertilizers depend upon their chemical properties.
4. Some time it forms complexes with other fertilizers.	4. Proper care taking while mixing avoids further problems.

Difference between organic farming and inorganic farming

Organic farming	*Inorganic farming*
1. It uses only organic fertilizers.	1. It uses only inorganic or both fertilizers.
2. Its cost may be higher than the inorganic farming.	2. Its cost may be lower than the organic fertilizers.
3. Organic farming is basically based on utilization of naturally available resources for the cultivation of farm.	3. Synthesized or manufactured fertilizers generally used for cultivation of various crops.
4. First dew years it gives fewer yields, but after that is gives consistent crop yield.	4. This type of cultivation gives higher yield throughout the cultivation period if other factors does not hamper.
5. Organic farming gives all nutrients through the natural resources. Therefore, their cost will be less than the manufactured fertilizers.	5. Inorganic fertilizers are various in types so they can provide one or more nutrients to the crop at one time but their cost may be higher.
6. In built resistance in the crop can be developed by using natural resources.	6. Insect, pests can become resistant against the pesticides, fungicides or insecticides so cost may be higher.
7. Pesticides, insecticidal, fungicidal residual effect will be less.	7. The residual effect of pecticides, insecticides, fungicides will be higher.
8. Cost of final product will be higher than the inorganic crop.	8. The cost of the product will be less than the organic farming.
9. Taste and quality of organic product is better than the inorganic product.	9. Taste and quality may be little bit lower due to the chemical effect on the product.
10. Demand of the food product fluffing is very difficult by this method.	10. Demand of the food product fulfilling by this method is very easy by enhancing the bumper yield.
11. All farm produce can be converted in to organic manure and that can be efficiently utilized for farm.	11. Industrial products will increase the cost of product.
12. Chemical hazard or fear will not occur in this system.	12. Human being has fear about chemical hazards in the food product.

5

REASONING

Different critical stages of crops for irrigation

Cereals are highly sensitive to soil moisture stress during panicle initiation and flowering. In the case of rice crop, highest yield reduction (70%) occurs due to stress at booting (reduction division) stage followed by that at early heading (64%). Soil moisture stress at crown root initiation stage of dwarf wheat leads to about 40% reduction in grain yield, while that at tillering reduces the yield by about 25%. Legumes are sensitive to stress at flowering and pod development stages.

Crop	Moisture sensitive periods
Rice	Panicle initiation, heading and flowering
Sorghum	Booting, flowering, milky and dough stages
Maize	Teaseling, silking and early grain formation
Pearl millet	Heading and flowering
Finger millet	Panicle initiation and flowering
Wheat	Crown-root initiation, shooting and heading
Barley	End of the shooting and heading
Groundnut	Rapid flowering peg penetration and early pod development
Sesame	Flowering to maturity
Sunflower	Flower bud initiation, head initiation, flowering and milky stages
Safflower	From rosette to flowering
Soybean	Flowering and seed formation
Cotton	Flowering and boll development
Sugarcane	Formative phase, particularly during tillering
Tobacco	Entire growth period
Chillies	Flowering
Potato	Tuber initiation to tuber maturity
Onion	Bulb formation to maturity
Tomato	From the commencement of fruit set

Contd...

Crop	Moisture sensitive periods
Peas	Flowering and pod development
Cabbage	Head initiation until becoming firm
Citrus	Flowering, fruit setting, fruit growth
Banana	Early vegetative period, flowering and fruit formation

All the stages of growth are equally sensitive to soil moisture stress for crops where vegetative parts are of economic importance. Total growth and yield of perennial plants are the summation of effects of stress at each growth stage. However, adequate water supply is essential at flower bud initiation, flowering and fruit set. Flower bud formation, however, increases due to restricted water supply prior to flower bud initiation in the case of citrus and mango. For realizing maximum benefit from the scarce irrigation water, irrigations are to be scheduled at moisture sensitive periods by withholding irrigations at other periods of lesser sensitivity. Such irrigation schedules along with improved management practices increase the water use efficiency in crop production.

Measurement of Irrigation Water

Water is the most valuable asset of irrigated agriculture. Efficient use of water for irrigation depends largely on measurement of water. Measurement reduces excessive waste and allows the water to be distributed among users according to their needs and rights. Information concerning the relationship between water, soils and plants cannot be utilized in irrigation practice without the measurement of water. Therefore, accuracy of water measurement is of primary importance in the operation of water distribution system.

Different units are used to express the water irrigated to a field. Under rest condition - *i.e.,* in ponds, tanks, reservoirs, water is measured in liters, hectare-centimeter, and cubic meter. When the water is in motion in rivers, canals, pipelines, field channels and cannel structures, the measurement of water is done in unit of rate of flow, *i.e.,* litres/sec, liters/hour, cubic meter per second, hectare- centimeter per day etc.

- *Litre*: A volume equal to one cubic decimeter of 1/1000 cubic meter.

- *Cubic* **meter**: A volume equal to that of a cube 1 meter long, 1 meter wide and 1 meter deep (1 cubic meter = 1000 litres).

- *Hectare-centimeter*: A volume necessary to cover an area of 1 hectare (10,000 sq m) to a depth of 1 centimeter (1 hectare centimeter = 100 cubic meters= 1000,000 litres).

- *Liter per* **second**: A continuous flow amounting to 1 liter passing through a point each second (generally used to denote the discharge of a pump, small stream or pipeline).

- ***Cubic* meter *per* second**: A flow of water equivalent to a stream 1 meter wide and 1 meter deep flowing at a velocity of 1 meter per second.

Weed and their Classification

Recently, weed is defined as a plant that originated under a natural environment and in response to imposed and natural environments, evolved and continues to do so as an interfering associate with our desired plants and activities. In this context it is well said that while all weeds are unwanted plants, all unwanted plants may not be weeds.

Despite the good intensions of the above accepted definitions of weeds, for all intents and purposes, about 30,000 plant species have been identified as definite weeds in the world infesting croplands, water bodies, wood lands, gardens, orchards, air fields, utility rights-of way, etc.

Weediness is defined in which there is an abundance of weeds. Depending on their migration into croplands, association with specific crop, problems created, survival, etc. weeds are named differently.

- **Parasitic weed:** A weed that depends partially or fully for its growth on its host plant.
- **Obligate weed:** A cropland weed incapable of surviving in a wild community.
- **Objectionable weed:** A problem weed, whose seed once mixed with crop seed, is extremely difficult to separate.
- **Noxious weed:** An undesirable troublesome weed difficult to control.
- **Facultative weed:** Weed of wild community origin, escaping sometimes to cropland.
- **Alien weed:** Weed not native of a country.
- **Associated weed:** Non-parasitic weed association with a specific crop.
- **Satellite weed:** A weed that has become an integral part of a crop ecosystem.

Weeds have been part of agriculture scene since man started cultivating crop plants and weed management is an integral part of crop production with the introduction of improved cultivars. Weed management is an approach in which weed prevention and weed control have companion roles. It implies a systems approach in which all available tools are used to reduce the propagate seed back, prevent weed emergence and minimize competition from weeds growing with desired plants. Thus, weed management has both immediate and long term objectives. This approach also implies a consideration in a boarder context of their interactions with production practices.

Instruments Used for Measuring Different Weather Parameters in Observatories and Laboratories

Instrument	Weather parameter
Maximum thermometer	Highest temperature attained during a day
Minimum thermometer	Lowest temperature reached during a day
Thermograph	Air temperature continuously
Thermocouples	Soil temperature
Soil thermometer	Soil temperature
Aneroid: barometer	Atmospheric pressure
Barograph	Atmospheric pressure continuously
Ordinary rain gauge	Amount of rain fall
Self-recording rain gauge	Amount of' rainfall/Intensity of rainfall
Psychrometer/hygrometer	Relative humidity
Assmann psychrometer	Relative humidity in crop canopy
Hair hygrometer	Relative humidity inside the room
Hygrograph	Relative humidity continuously
Wind vane	Wind direction
Anemometer	Wind velocity
USWB Class A open pan	Evaporation
Heliograph	Duration of bright sunshine hours
Bellani pyranometer	Total incoming radiation
Pyrgeometer	Long wave radiation
Spectroradiometer	Solar radiation in narrow wavebands
Kipp radiometer	Diffused radiation
Eppley spectral pyranometer	Photosynthetically active radiation (PAR)
Pyradiometer	Both long and short wave radiation
Net radiometer	Net radiation
Quantum sensor	Quantum content in radiation
Spectrophotometer	Wavelength of light

Deficiency Conditions of Nutrients

Nutrient	Condition inducing deficiency
B	Sandy soils, acidic leached soils, alkaline soils with free lime
Ca	Acidic, alkali. or sodic soils
Fe	Calcareous soils, soils with high P, Mn, Cu or Zn, high rate of liming

Contd...

Nutrient	Condition inducing deficiency
K	Sandy, organic, leached and eroded soils, high liming, intensive cropping
Mg	Similar to calcium
Mn	Calcareous silt and clays, high organic matter
Mo	Highly podzolised soils well drained calcareous soils
N	Low soil organic matter, leaching
P	Acidic, organic, leached and calcareous soils, high rate of liming
S	Low soil organic matter, use of high analysis fertilizers with low S
Zn	Highly leached acidic soils, calcareous soils, high soil levels of Ca, Mg and P

Functions of Various Nutrients

Nutrient	Function(s)
Boron	It is essential for development and growth of new cells in plant meristem. It is necessary for nodule formation in legumes. It is associated with translocation of sugars, starches, nitrogen and phosphorus.
Calcium	It is involved in cell division and plays a major role in maintenance of membrane integrity.
Carbon	Basic molecular component of carbohydrates, proteins, lipids and nucleic acids.
Chlorine	Essential for photosynthesis and as an activator of enzymes involved in splitting water. Associated with osmoregulation of plants growing on saline soils.
Copper	It acts as electron carrier in enzymes associated with oxidation reduction reactions. It has indirect effect on nodule formation.
Hydrogen	Plays a central role in plant metabolism. Important in ionic balance and as main reducing agent plays a key role in energy relations of cells.
Iron	An. essential component of many hemo and non-hemo Fe enzymes and carriers, including cytochromes (respiratory electron carriers) and the ferredoxins. The latter are involved in key metabolic functions such as N fixation, photosynthesis and electron transfer.
Magnesium	Component of chlorophyll and a cofactor for many enzymatic reactions. It is a structural component in ribosomes.
Manganese	Involved in oxygen evolving system of photosynthesis. It can substitute for magnesium in many of the phosphorylating and group transfer reactions.lt influences auxin levels in plants.
Molybdenum	It is an essential component of enzyme nitrate reductase in plants. It is also. a structural component of nitrogenase associated with nitrogen fixation in legumes.
Nitrogen	It is a component of many important organic compounds ranging from protein to nucleic acids. It is an integral part of chlorophyll, which is the primary absorber of light energy needed for photosynthesis. It imparts green colour to plants.

Contd...

Nutrient	Function(s)
Oxygen	It is somewhat like carbon in that it occurs in virtually all organic compounds of living organisms.
Phosphorus	Central role in energy transfer and protein metabolism. It is an important structural component of many bio-chemicals including nucleic acids. DNA and RNA are associated with control of hereditary processes. It is also associated with increased root growth and early maturity of crops.
Potassium	Helps in osmotic and ionic regulation. It functions as cofactor or activator for many enzymes of carbohydrate and protein metabolism. Imparts disease resistance in cereals and drought resistance in many crops.
Sulphur	Like phosphorus, it is involved in plant' cell energetic. It is associated with chlorophyll formation and sulphur containing amino acids.
Zinc	It is a constituent of several enzyme systems regulating various metabolic reactions.

Origins of Various Crop Plants

Crop	Origin of the Crop
Cereals	
Rice	Indo-Burma (Indo-Myanmar), Jaypur (Orissa) Secondary
Wheat	South West Asia
Maize	Mexico
Jowar	Ethiopia and Sudan (Africa)
Bajra	Africa
Barley	Abyssinia (Ethiopia)
Pulses	
Gram	South West Asia
Field Pea	Mediterranean region
Arhar	Africa (Nile river and Angola)
Mung/Urd	India (Central Asia)
Cowpea	Central Africa
Soybean	China
Oilseeds	
Groundnut	Brazil
Sesame	South West Africa
Castor	Ethopia
Rai	China
Rapeseed	Afganistan, Pakistan and India
Sunflower	Mexico and South USA (Central America)

Contd...

Crop	Origin of the Crop
Other Crops	
Cotton	India
Corchorus olitorius	Africa
Corchorus capsularis	Indo-Burma (Indo-Myanmar)
Tropical Cane	Oceania (New Guinea) North
Indian Cane (*Saccharum spontaneum*) (Kans)	North Eastem India
Sugarbeet	Mediterranean region
Potato	South Africa (Peru)
Tobacco	Mexico and Central America

Classification of Various Types of Fertilizers

1. **Straight Fertilizers:** Such fertilizers have declarable content of only one major nutrient, *e.g.,* urea, ammonium sulphate.

2. **Binary Fertilizer:** Contains two major nutrients *e.g.,* Potassium Nitrate.

3. **Ternary Fertilizer:** Contains three major nutrients *e.g.,* Ammonium potassium phosphate.

4. **Compound/Complex Fertilizer:** Such fertilizer has a declarable content of at least two of the major nutrients obtained chemically and generally granular in form, *e.g.,* Nitrophosphate, Ammonium-phosphate and Diammonium phosphate (DAP).

5. **Mixed Fertilizer:** Individual or straight fertilizer materials are blended together physically to permit application in the field in one operation. Such fertilizers supply two or three major nutrients in a definite proportion of grade, *e.g.,* Nitrophosphate with potash 15:15:15 of NPK.

6. **Complete Fertilizer:** Having all the three primary major nutrients, *viz.,* N, P and K.

7. **Incomplete Fertilizer:** Containing any two primary nutrients.

8. **Low-analysis Fertilizer:** Having less than 25% of the primary nutrients *e.g.,* SSP (16% P_2O_5), sodium nitrate (16% N).

9. **High-analysis Fertilizer:** Having more than 25% of the total primary nutrient content, *e.g.,* urea, anhydrous ammonia (82.2% N), Ammonium phosphate (20% N + 20% P_2O_5) and DAP (18% N + 46% P_2O_5).

10. **Fertilizer grade:** Refers to the guaranteed analysis of its plant nutrients. It is the minimum guarantee of the plant nutrient contents in the term of N, available P_2O_5 and K_2O, *e.g.,* 6:24:24.

11. **Fertilizer Ratio:** refers to the relative percentage of N, P_2O_5 and K_2O, i.e., 1:4:4 if fertilizer grade is 6:24:24.

Materials used in Manufacturing of Fertilizers (Mixed)

1. **Supplier of plant nutrient:** Straight fertilizers are used for these purposes.

2. **Conditioners:** To check absorbing moisture and making one, conditioners like straw, groundnut husk, peat soil etc. are used they just reduce caking and applied in drilling conditions and these conditioners are of low organic materials.

3. **Neutraliser of acidity or basicity:** Dolomitic limestone is used to reduce residual acidity. Most fertilizers leave residual acidity or basically.

4. **Filler materials or make weight materials:** Sand, soil, earth, coalash, charcoal such waste materials are added to make up the difference between the weight of the added fertilizers required to supply the plant nutrients and the desired quantity of the fertilizer mixture.

Land Capability Classification

Soil survey maps and reports are the basis for systems of land capability classification. Land capability classification is grouping of soils into different classes according to their capability for intensive use and treatments required for sustained use. It emphasizes the need for using the land only for what it is suited best to realize optimum returns without land degradation. Land capability classification system developed by USDA is useful for agriculture. Eight land capability classes are recognized in this system of classification.

Class I Has few limitations on their use. They are deep, well drained and nearly level; they are fertile or responsive to fertilizer application. A variety of crops can be grown intensively with recommended management practices.

Class II Soils have few limitations such as gentle slope, moderate erosion problem, inadequate depth, slight saline and alkaline problem and relatively restricted drainage. Less intensive cropping system must be followed. Simple management practices such as contour cultivation can maintain the soil for crop production.

Class III Soils can be used for crop production with special conservation practices like terracing. Soil erosion, shallow water permeability, low moisture retentively, moderate salinity and low fertility are the limitations in their use. Smothering crops such as legumes are more ideal than row crops.

Class IV Can be used for cultivation with severe limitations on the choice of crops. Steep slopes, severe erosion, shallow soil depth, low moisture retentivity, salinity or alkalinity restricts profitable crop production. These lands should be used for close growing crops or grasses with soil conservation practices.

Class V Is generally not suitable for grain crops due to limitations such as rocky soil, ponded areas with no drainage facility, etc. Pastures can be improved on this class of land are not suitable even for growing grasses. Limitations are the same as those for class V land, but they are more rigid. Their use may be restricted to woodland or wild life.

Class VII Soils are severely limited even for growing grasses and forest trees. They are steep rough stony soils of extremely shallow depth.

Class VIII Lands are not suitable even for forest trees as they are steep rough stony mountains. Land use is restricted to recreation, wild life, etc.

In each of the land capability classes there are subclasses that have some kind of dominant limitations for agricultural use. The four kinds of limitations in the subclasses are risks of erosion (e), wetness (w), soil depth (s) and climate (c). Thus soil class III (e) indicates risk of erosion in Class III.

Different Horizons of Soil

The layers resulting for soil forming processes are grouped under five heads: O, A, E, B, and C and all these horizons in combination are called Master Horizons.

1. **'O' horizon:** Organic horizons above mineral soil, occurring commonly in forest areas. Such 'O' horizon is visible in virgin soil and absent in arable soils.

2. **'A' horizon:** Top most mineral horizon, containing a strong admixture of humified organic matter which tends to impart a darker colour than that of the lower horizon.

3. **'E' or 'A$_2$' horizon:** Horizon of Maximum Eluviation of clay, Fe and Al oxides and a corresponding concentration of resistant minerals such as quartz in sand.

4. **AB/EB horizon:** Transition layer between A (or E) and B with properties more nearly like those of A (or E) than of underlying B (formally called A$_3$). Sometimes it is absent.

5. **'B' horizon:** Horizon of maximum accumulation (Illuviation) of materials such as Fe and Al oxides and silicate clays (Illuvial horizon) and also CaCO$_3$, CaSO4 and other salts in arid zones. Organic matter content is generally higher than that of E.

6. **'C' horizon:** Unconsolidated mineral underlying the solum (A and B) zone of least weathering, accumulation of Ca, Mg carbonates, cementation, sometimes high bulk density.

Agro-forestry

Agro forestry may be defined as an integrated self-sustained land management system, which involves deliberate introduction/retention of woody components with agricultural crops including pasture/livestock, simultaneously or sequentially on the same unit of land, meeting the ecological and socio-economic needs of people. It is also defined as a collective name of land use systems and technologies where woody perennials are deliberately used from the same land management units as agricultural crops and/or animals in some form of temporal sequence. In agro forestry systems, there is both ecological and economic interaction between different components.

Based on the kind of associated agricultural products, major function of the tree component, spatial arrangement of trees and duration of combination, several systems have been identified. In India, agro forestry systems are classified as:

Agri-silviculture	:	Crops + trees
Silvi-pasture	:	Trees + pasture
Agri-silivi-pasture	:	Crops + trees + pasture/animals
Horti-pasture	:	Fruit trees + MPTS + pasture/animals
Agri-horti-silviculture	:	Crops + fruit trees + MPTS

Different Recommendations for Dry Land Agriculture

Recommendations for Dry Farming Areas

The research programmes of the All India Co-ordinated Research Project for Dry Land Agriculture have concluded into certain justified claims of research achievements which have made the following distinct recommendations:

1. Bunding across the slope and levelling the land should be done before onset of monsoon.

2. Deep summer ploughing should be followed by surface tillage during monsoon months and also rest of the year.

3. Application of organic manures like F.Y.M., compost @ 15-20 tonnes/ha or green manuring should be done. These manures should be applied about 20-25 days before sowing and should be well mixed in soil.

4. Fertilizers should be basal placed at a depth of 7.5 to 10 cm in the soil and the seeds should be sown in the same furrows about 3 cm above the

fertilizers. This is important especially during winter season. The nitrogen (25-50% of the total) should be top-dressed by side or band placement method at about 10-15 cm apart the crop rows which should be done soon after the rains but if there is no moisture in the soil the nitrogen should be sprayed over the foliage with urea solution containing 5-15% nitrogen. Zn and sulphur should be applied as basal, if needed.

5. Soil applications of BHC (10%) dust @ 25-30 kg/ha for termites and Thimet 20G @ 15 kg/ha for white grub should be done and these chemicals must be mixed well in the soil at the time of final field preparation.

6. Selection of suitable crops and their varieties, as mentioned in genetic approaches of the text, must be done.

7. Seeds must be treated with a suitable fungicide and that of legumes with a bacterial culture before sowing. Soaking seeds in plain water for *rabi* sowing helps in getting higher germination, better seedling vigour and an early maturity by a week's time.

8. Proper crop rotation should be followed which should preferably have at least one legume crop every year.

9. For better seed-soil-moisture contact thorough soil compaction should be done by running a planker or a roller, especially after *rabi* planting.

10. At the event of total crop failure during *kharif* season a suitable catch crop like urd T.9 or toria etc. should be sown.

11. Inter-cropping of oil seeds and pulses should be done with jowar, bajra and maize crop for the purpose of making best use of soil and inter row moisture harvesting. This also increases crop productivity/unit area/ unit time.

12. Line sowing by drilling the seed at a depth of 7.5 to 10 cm or even more depending upon the situation should be practiced because it helps in better seed germination, high drought resistance, etc. line sowing also facilitates the agricultural operations. On terraces strip cropping should be practiced.

13. Twenty five per cent more than the recommended seed rate should be used and thinning of excess plants should be done about 2-3 weeks after germination. This helps in stabilizing the required plant population and thereby in getting better yield.

14. Proper weed management practices should be followed by adopting integrated weed control measures like one hand weeding about 20-25 days after sowing, then one hoeing about 5-10 days later followed by use of appropriate herbicide for an effective weed control.

15. Mulching should be done by providing frequent inter culture and pulverizing the soil. If intercultural operations are not possible then use of artificial mulches like covering the surface by organic refuse (tree leaves, uprooted weeds, sugarcane leaves, saw dust or polythene sheets) are used to check the evaporation loss of water from the soil.

16. Water harvesting between the rows should be done by growing pulse crops and run-off water should be collected in some nearby located pond which may be used for recycling in the form of protective or life saving irrigation to the crop.

17. An efficient plant protection measure should be adopted to protect the plants from various insect/pest and disease damage.

18. Crops should be harvested at physiological maturity so that the following or succeeding crop may be sown slightly earlier than the scheduled time and best use of rain water or residual moisture may be made for crop production.

19. Crops like cotton, chillies should be sprayed with CCC or Cycocel and groundnut should be sprayed with Plano fix for modified growth, higher drought resistance and better yield.

Intercropping for Dry Lands

Intercropping recommended for different regions of India under dryland conditions:

Region	Intercropping
Samba (Jammu)	1. Castor + black gram/sesame/cowpea/sunflower 2. Maize + cowpea (fodder) 3. Maize + green gram
Hissar (Haryana)	Pearl millet + cluster been/green gram
Dehradun (U.P.)	Pigeon pea + soybean, Maize + pigeon pea
Agra (U.P.)	1. Pearl millet + black gram/green gram 2. Pigeon pea + green gram/pearl millet
Varanasi (U.P.)	Pigeon pea + maize / black gram/millet
Ranchi (Bihar)	1. Barley + rapeseed 2. Wheat + Bengal gram/rapeseed/lentil
Bhubaneswar (Odisha)	Pigeon pea + black gram/sorghum/groundnut
Rewa region (M.P.)	Pigeonpea + sorghum + finger millet black gram/sunflower
Jhansi (M.P.)	1. Sorghum + black gram/green gram/soybean' 2. Wheat + Bengal gram/mustard 3. Barley + Bengal gram/mustard 4. Bengal gram + linseed/mustard

Contd...

Region	Intercropping
Indore (M.P.)	Pigeon pea + sorghum, Cotton + soybean
Udaipur (Rajasthan)	Sorghum + black gramPearl millet + black gramSunflower + black gram
Anand (Gujarat)	Pearl millet + red gram/castor/sunflower
Akola (Maharashtra)	1. Pigeon pea + black gram/green gram/sorghum/groundnut 2. Cotton + sunflower/black gram/groundnut
Sholapur (Maharashtra)	1. Pigeon pea + pearl millet/sorghum/black gram 2. Cotton + sunflower/groundnut
Anantapur (A. P)	Pigeon pea + green gram/pearl mill/sorghum/foxtail millet/groundnut/sunflower.
Bijapur (A.P)	Cotton + ground nut/sorghum/pearl millet/green gram
Mysore region (Kamataka)	1. Pigeon pea + pearl millet 2. Cotton + pearl millet 3. Groundnut + cotton
Kovilpatti (T.N.)	1. Cotton + black gram/green gram 2. Sorghum + black gram/cowpea/Dolichos lab lab 3. Pearl millet + green gram/cluster bean/cowpea/Dolichos lab lab

Classification of Salinity of Water

C_1 *Low Saline Water.* If electrical conductivity is less than 0.25 dS/m, the irrigation water is classified as low salinity water. It can be used for irrigation on all soils and on most crops but leaching is required in case of extremely low permeable soil.

C_2 *Medium Saline Water.* It has EC between 0.25 to 0.75 dS/m. This water can be safely used for crops with moderate salt tolerance. The soil should have moderate level of permeability and leaching to avoid accumulation of salts.

C_3 *High Saline Water:* Water within ranges of 0.75 to 2.25 dS/m, is called high salinity water. This water can be used for salt tolerant crops by providing good drainage and by practicing good management practices for salinity control.

C_4 *Very High Saline Water.* If EC is, more than 2.25 dS/m the water is classified as very high salinity water. It is not suitable for irrigation under ordinary conditions but may be used occasionally if the soil is permeable by providing adequate drainage.

Classification of Sodium Levels of Water

S_1 *Low Sodium Water.* It can be used for irrigation on almost all soils with little danger of the development of sodium problem.

S₂ *Medium Sodium Water:* It can only be used on soils of coarse texture and with lot of organic matter. It produces harmful effect if applied to the soils of fine texture and low permeability.

S₃ *High Sodium Water:* It may produce troublesome sodium problems in most soils and requires good management like providing good drainage, high leaching and addition of organic matter. If there is plenty of gypsum in the soil, no problem develops for some time. If gypsum is not decrease sodium hazard appreciably in case this water is to be applied to the soil.

S₄ *Very High Sodium Water:* It is not fit for irrigation except at low and medium salinity. Gypsum has to be added to the soil to decrease sodium hazard appreciably in case this water is to be applied to the soil.

Most Serious Weeds in the World

Common name	Scientific name
Smooth pig wood	*Amaranthus hybridus*
Spiny amaranth	*Amaranthus spinosus*
Wild oat	*Avena fatua*
Common lambsquarters	*Chenopodium album*
Field bind weed	*Convolvulus arvensis*
Bermuda grass	*Cynodon dactylon*
Yellow nut sedge	*Cyperus esculentus*
Purple nut sedge	*Cyperus rotundus*
Crabgrass	*Digitaria sanguinalis*
Jungle rice	*Echinochloa colonum*
Barnyard grass	*Echinochloa crusgalli*
Water hyacinth	*Eichhornia crassipes*
Goose grass	*Elesine indica*
Cogon grass	*Imperrata cylindrical*
Sour paspalum	*Paspalum conjugatum*
Common purslane	*Portulaca oleracea*
Itch grass	*Rottboellia exaltata*
Johnson grass	*Sorghum halepense*

Classification of Herbicides

On the basis of Chemical Structure

(a) Inorganic Herbicides: First chemical used, Arsenic, Sod Sulphuric acid, Sodium arsenate, Sod Chlorate, Borax, Copper Sulphate, Copper nitrate, etc.

(b) **Organic Herbicides:** 16 or 17 groups:

Group	Herbicides
Aliphatic	Datapon, TCA, Aerolein, CH_3 Br.
Amides and Acetamides	All-chlor (like Butachlor, Akhlor (Lasso), Propachlor, Propanil.
Baenjoics	2,3,6-TBA, Oicamba, Tricamba, Chloramben.
Bipyridiliums	Paraquai, Oiquat (contact).
Carbamates	Propham, barban, dichlormate.
Thiocarbamates	Butylate, Thiobencarb or benthiocarb.
Dithio carbamates	CDEC, Metham.
Nitriles	Bromoxynil, Dichlobenil.
Dinitro anilines	Fluchloralin (Basalin), Pandi methalin, Trifluralin, Nitralin.
Phenols	Diniseb, DNOC, PCP.
Phenoxys	2, 4-0, 2, 4, 5-T, MCPA, MCPB, 2,4-6, Dichlorprop.
Triazines	Atrazine, Propazine, Simazine (soil applied).
Ureas	All-ron (like Diuron, Monuron, Isoproturon).
Uracils	Bromacil, Terbacit.
Diphenyl ethers	Nitrofen (Toke-25).
Others	Picloram, pyrazon.

Relative Persistence of Some Common Herbicides in Soil when Applied at Recommended Rates for Weed Control

Less than 3 months	3-6 months	More than 6 months
Aminotriazo	Chlorbromuron	Atrazine
Chlorpropham	Diallate	Bromacil
Cunazine	Dinitroamine	Dichlobenil
Dalapon	Isoproturon	Diuron
Metoxuron	linuron	Methazone
Prometryn	Methabenzthiazuran	Metribuzin
Propachlor	Metabromuron	Simazine
Propham	Propyzamide	Terbacil
Terbutryn	Pyrazon	Trifluralin

Determination of Crop Water Requirement

There are several methods of determining crop water requirement. However, there methods depend on the desired level of accuracy, availability of equipment and technical know how.

1. Transpiration ratio method
2. Depth-interval yield approach method
3. Soil moisture depletion studies
4. Field experimentation
5. Climatological approaches
6. Drum-culture technique for lowland rice
7. Water balance method

Drainage Systems

1. Surface drainage	2. Sub-surface drainage
3. Interception drains	4. Pipe or tile drainage
5. Diversion drains	6. Gridiron
7. Bedding system	8. Herring bone
9. Random system	10. Parallel system
11. Drainage well	

Principle Gases Present in the Air

Sr. No.	Gas	Volume (%)	Weight (%)
1.	Nitrogen	78.088	75.5270
2.	Oxygen	20.949	23.1340
3.	Argon	0.930	1.2820
4.	Carbon dioxide	0.030	0.0456
	Total	**99.997**	**99.997**

Criteria of Drought

Production above 75% of normal	: No drought
Production 50 to 75% of normal	: Moderate drought
Production 25 to 50% of normal	: Severe drought
Production less than 22% of normal	: Disastrous drought

Methods of Fertilizer Application

1. Broadcasting
2. Top dressing
3. Placement
4. Band placement
5. Pellet placement
6. Side dressing
7. Plough-sole placement
8. Deep placement
9. Localized placement
10. Contact placement
11. Drill placement

Oil Percentage of Some Medicinal Plant used in IPM

Medicinal plant	Oil percentage
Kusum (*Schelechera oleosa*)	57 – 60
Mahua (*Madhuca indica*)	33 – 43
Neem (*Azadiracta indica*)	40 – 50
Persian lilac (*Melia azedarch*)	25 – 30
Pongam or Karanj (*Pongamia glabra*)	27 – 30
Sal (*Shorea robusta*)	12 – 13

Various Field Crops Varieties and Hybrids

S. No.	Name of the crop	Varieties/Hybrids
1.	Barley	BHS46, Dolma, Himani, BG 105, BG 25, BH39, Clipper, DL36, DL 70, DL85, HBL3, IB65, K18, Jyoti, K226, K287, Amber Azad, K 141, K 169, K252, P 103, PL56, Ratna, Bilara, K19, K257, Rajkiran, RD 117, RD 137, RD31, RD57, RDB 1, RS6, Vijay.
2.	Barnyard millet	KL 29, Kanchan, VL 1, IPS 149, UPT 8, K 1, T 46, PC 49, IPM 100, IPM 148, IPM 151, PRB 9403.
3.	Black gram	MASH-1-1, T 9, Kulu Mash 1, Pant U 19, VB3, T9, Navin, Pant U 19, PDU 1, Narendra Urd1, PDU 88-108, WBU 108, MASH 1-1, T9, G75, Pant U 19, PDU 1, Pant U 35, WBU 108, Pant U 30, Sindkheda1-1, Zandewal, Jawahar Urd 2, Jawahar Urd 3, TPU4, TAU 2, VB3, Khargone 3, Co2, KM2, TMV 1, Co5, ADT3, Sarala, LBG 17, LBG402, Pant U 30, LBG 20, Vamban 1, ADT4, ADT5, VB3, WBG26, TPU 4, TPU 1, BDU 1, PU 9503-8, PU 9205-11-3,T 9, TPH 4, PKV 15, AKV 4.
4.	Caster	Aruna, Bhagya, Sowbhagya, RC8, SA2, TMV5, Co 1, T3, T4, GAUCH 1, GCH4, GCH2, DCS9, SKI 73, DCH 32, GC 2, GCH5, PCS 4, DCH 177, TMVCH 1, JAB 665

Contd...

S. No.	Name of the crop	Varieties/Hybrids
5.	Chickpea	**Kabuli**: L550, ICCC 32, Pusa 267, ICCV 2.**Desi:** C 235, Radhey, K 850, JG 315, Pusa 256, Phule G 5, Avrodhi, PBG 1, Chana No. 1, JG 74, RSG 59, ICCV 10, Pusa 372, Vijay, GVG 663, KWR 108, Pusa 391, GNG 469, CSG 8962, DCP 92-3, Viswas, Phule G 12, Vishal, Virat, Vihar, Digvijay, BDN 9-3, BDNG 797, PKV 2, Green Chafa, D 8, Gulak 1, ICCV 2, 10, AKG 46, Jaki 9218, Saki 9516, Gulab 1, PKV Gulabi 2.
6.	Cotton	F 286 LH 886, F 414 LH 900, F 505 LH 1134, F 846 F 1084, F 1378, LH 1556, H 777, HS 45, HS6, H1098, G. agethi, HS 6, RST 9, B. Nerma, RS 875, Vikas, Kandwa 3, Vikram, KC 94-2, PKV 081, LRK 516, LRA 5166, LRK 516, CNH 36, Rajat, G. Cot 12, LRK 516, G. Cot 14, CNH 36, G. Cot 16, LRA 5166, NA 920, LK 861, L 389, Kanchan, Sharda, JK 119, Abhadita, MCU 5, MCJ5VT, MCU 7, MCU 9, MCU 10, MCU11, LRA5166, LRK516, Suvin (Anjali), Surabhi,
		Phule 388, NHH 44, Hybrid 10, Phule 492, LRH 5166, JLH 168, Y 1, Ganga, NNH 44, NH 545, PKV Rajat, AKH 8828, DHY 286, AKH 081, AKA 5, 7, 8, 8401.
7.	Cowpea	C 152, V 240, FS 68, T 2 Pusa Sawani, Pusa Dophasli, RC 29, JC 10, Cowpea 74, Pusa Phalguni, Amba, RS 9 Amba, T 5269, C 152, RC 19, T 2, Gomti, Pusa Sawani, Cowpea 74, V 240, FGS 1, T 5269, FOS 42-1, K 11, C 152, V 240, Amba, No. 21, GC 1, GC 2, K 14, GC 5-19-4-1, C 152, V 240, Amba, PTB 1, Krishnamani, S 488, Co 1, Co 2, S 288, JC 5, Co 3, Paiyur 1, Phule Pandhari, C 152, Konkan Sadabahar, Konkansafed, C 152, V 16, RC 19.
8.	Field pea	Type 163, Rachna, DMR 11, Pant P 5, HUP 2, JP 885, HFP 4, KFP 103, DMR 7, HFP 8909, KPMR 144-1, Sel 93, Sel 82, T 163, KPMR 10, Koparkeda, Rachana.
9.	Finger millet	PES 400, VL 124, KM 65, VL 146, PES 176, Co 13, TRY 1, Indef 5, EC 4840, GPU 28, MR 1, MR 2, MR 374, Gautami, Padmavathi, RAU 8, Indef 1, A 404, RM 2, VL 149, GN 3.
10.	Foxtail millet	K 3, AK 132-1, Krishnadevraya 2, Gavari (SR 11), TNAU 43, ISC 201.
11.	French bean	PDR 14, HUR 15, HUR 137, VL 63.
12.	Green gram	Pusa Baisaki , PS 16, K 851, PDM 54, Amrit, Sunaina, Panna, Sonali, ML 337, Pant Mung 1, PDM 54, Narendra Mung1, PDM 84-143, PDM 84-139, PDM 11, G65, ML5, ML1, ML 131, SML 32, ML 337, S9, Pant Mung 3, ML 267, Pusa 105, RMG62, MUM2, Asha, MH88-111, Jawahar 45, Kopergaon, ML 131, Jalgaon 781, TAP 7, Pant Mung 2, Mung Guj. 1, Sabarmati, ML 337, PDM 11, Pusa 105, BM4, Phule M 2, HUM 1, TARM 1, Kondaveedu, KM 1, KM2, Co4, Jyoti, Sujata, Dhauli, Ratila sel, Co2, Co 3, Paiyur 1, PDM 54, Vamban 1, PDM 84-143, ADT3, Co5, Pusa 9072, HUM 1, TARM 1, Vibhav, Phule M 2, PM 9339, BM 4,

Contd...

S. No.	Name of the crop	Varieties/Hybrids
		Kopargaon, BPMR 145, BM 2002-1, TARM 2, 18, PKV 8802, AKM 9910, AKM 9911.
13.	Groundnut,	JL 24, CO 1, Kisan, KRG 1, TG 17, GG 2, Jawan, CO 2, Dh 8, TG 3, ICGS 11, SG 84, VRI 1, Gimar, ICGS 44, Rg 141, VRI 2, ICGS 37, MH 4, ICGS 1, ICGS 10, VRI 3, RSHY 1, TAG 24, ICGV 86530, Tirupati 2, Konkan Gaurav, GG 3, Gangapuri, Kopargaon 1, TMV 11, MH 2, TMV 10, TG 1, Kadiri 2, Kadiri 3, BG 1, BG 2, M 197, Chitra, Kaushal, UF 70-1-3, MA 16, ALR, BG 3, RS 138, ICGS 76, ICGS 86325, BAU 13, DRG 17, DRG 12, M 145, S 230, Kadir 71-1, M 13, GAUG 10, Chandra, M 37, Phule Pragati, Phule Vyas, JL 220, TG 26, SB 11, M 13, TMV 10, TAG 24, ICGS 11, Koyana (B 95), Karad 4-H, AK 159.
14.	Jute	JRO 632 (Baisakhi Tossa), JRO 524 (Navin), JRO 7835, JRO 524, JRO 7835, JRO 632, JRO 524, JRO 7835, JRO 878 (Chaitali Tossa), T J40, (Mahadev), . KOM 62 (Rebati), JRO 524, JRO 3690 (Sabitri), JRO 632, JRO 3690, JRO 524, JRO 7835, JRO 878, JRC 212 (Sabuj Sona), JRC 321, (Sonali), JRC 7447 (Shyamali), UPC 94 (Reshma), Hybrid C (Padma), JRC 212, JRC 321, JRC 7447, KTC 1 (Rajendra Sada Pat), JRC 212, JRC 7447, UPC 94, JRC 4444 (Baldev), KC 1 (Jaydev).
15.	*Kodo* millet	IPS 147-1, JNK 364, Rewa, JK62, JK76, GPUK3, APK 1, GK2, KMV20 (Vamban), Kherapa.
16.	Kulti	K 42 (Man), D 40-1(Seena), Dapoli 1.
17.	Lentil	Pant L 406, Pant L 639, Malika, Lens 4076, LL 147, VL Masoor 4, Sehore 74-3, WBL 58, Sapna, JLS 1, Pant L 4, DPL 15, DPL 62.
18.	Little millet	Pajyur 1, Birsa Gandhi 1, TNAU 63.
19.	Maize/Corn	Ganga Hybrid Makka 1, Ganga Hybrid Makka 101, Ranjit Hybrid Makka, Deccan Hybrid Makka, VL54, Ganga Safed Hybrid 2 Makka, Histarch Hybrid Makka, Ganga Hybrid Makka 3, Himalayan Hybrid Makka 123, Sweet corn Madhuri, pop corn Amber Cpopcorn, Kiran, Panchganga, Pusa hybrid maka 1, 2, MMH 133, X 1123G, Deccan hybrid 1, 109, Prabhat, Karvirmaka, Manjri, New Jyoti, MMH 69, KH 9491, DMH 107, Y 1402K, JK 2492, Pro 311, Bio 9681, SSF 7374.
20.	Niger	No. 71, Gaudaguda Local, Ootacamund, No.5, GA 10, GA 5, RCR 17, IGP 76, Ootacamund, IGP 76, GA 5, IGP 76, RCR 317, Surabhi, Janki Himalini, R552.
21.	Pearl millet	HB 1, HB2, HB3, HB4, HB5, NHB3, NHB4, NHB5, PHB 10 (HB 6), PHB 14 (HB 7), GHB 1399, BJ 104, CJ 104, BK560, BD 111, COH12, CM46, GHB 27 (MH 28), BD 763, MBH 110, GBH 32 (MH 29), PHB 47, X5, HHB45, MBH 118, MH 179 (ICMH 451), MH 180, MH 182, HHB 50, BGH 50, MH 143, PUSA23, MH 312, HHB60, MBH 136

Contd...

S. No.	Name of the crop	Varieties/Hybrids
		(MH 123), MBH 149 (MH 249), HHB 67, VBH 67, MLBH 104, Eknad 301, HHB 68, ICMH 356, HS 1, PSB8, WC-375, Co6 RCB2, HC4, tCMS 7703, Co7, Sangam (RHR 1), PCB 15, ICMV221, Puaa266, CZ-IC 923, Pusa 23 (MH 169), Pusa 322, ICMH 451 (MH 179), ICMH 356, HHB 60, HHB 67, HHB 68, HHB 50, RHB 30, RHB 90, MH 605 (Pusa 605), MH 790, MH 782, HS 1, PSB8 WC-375 Co6, RCB2 HC4, ICMS 7703, Co7, Sangam (RHR 1), PCB 15, ICTP 8203, PusaSafed, ICMV 155, Raj 171, ICMV221, Puaa266, CZ-IC 923, RHRBH 8609, RHRBH 8924, ICTP 8203, JHB 558, MH 258, AHB 1666, AIMP 92901, PPC 6, PPC 2.
22.	Pigeonpea	Pusa Ageti, T21, UPAS 120, ICPL151, ICPL 85010, UPAS 120, AL15, Manak, Pusa 33, ICPL 151, Pussa 855, H 82-1, PPH 4, AL 201, T17, B 517, B7, Bahar, Pusa 9, OA 11, Birsa Arhar, BON 1, BON 2, JA3, Pusa 74, No. 148, PT 301, C11, PT 221, Hy 185, TI6, T 15-15, TAT 10, ICPL 87, ICPL8863, AKPH 4101, Pusa 33, ICPL 151, ICPHo8, JA 4, GUJ.tur 100, ICPL 87119, KM.7, BSMR 736, BSMR 175, Hy3A, HY3C, LRG 30, Gs 1, PT221, TI6, SA 1, ICPL87, ICPL 8863, Vamban 1, KM7, CoH 1, ICPL 332, Co6, ICPL84031, CoH2, TIB7, ICPH8, BDN2, C 11, ICPL 85010, ICPL 87, T-Vishaka, AKT 8811, Vipul, BSMR 853, BSMR 736, ICPL 87119, Badnapur 2, BDN 708, Konkan tur 1, C 11, No. 148, BDN 2, TAT 10, AKM 9911, AKM 9910, PKV 8802, TARM 2, 18, AKPH 2022, ICPL 8863, AKPH 4101, AKT 8811.
23.	Potato	Kufri Kisan, Kufri Kuber, Kufri Kumar, Kufri Kundan, Kufri Red, Kufri Safed, Kufri Neela, Kufri Sindhuri, Kufri Alankar, Kufri Chamatkar, Kufri Chandramukhi, Kufri Jeevan, Kufri Jyoti, Kufri Naveen, Neelamani, Sheetman, Kufri Muthu, Kufri Lauvkar, Kufri Dewa, Kufri Badshah, Kufri Bahar, Kufri Lalima, Kufri Sherpa, Kufri Megha, Kufri Ashoka, Kufri Jawahar, Kufri Sutlej, Kufri Pukhraj, Kufri Chipsona 1, Kufri Chipsona 2, Kufri Giriraj.
24.	Proso millet	Co 1, MS 1685, PV 196, MS 4848, Co 4, Nagarjuna, Bhawna.
25.	Rapeseed	Laha 101, BR 40, RT 11, Varuna (PP) , Patan 67 (Gujarat), Ourgamani, Rl 18, BS 2, B 70, BSH 1, Brown sarson, T 151, Patan sarson 66, YS Pb 24, Yellow sarson, M 27, T 9, BR 23, OK 1, PT 303, Shekhar, Vaibhava (RK 1418), Vardan (RK 1467), Rohini (KRV 24), Kranti (PR 15), Pusa Bold, Prakash, RH 30, Bhagirathi (RW 351) and Seeta (B 85) of mustard; Pusa Kalyani, Kosal of brown sarson, Benoy (B 9), 66-197-3, K 88 of yellow sarson, T 29, Agrani (B 54), Sangam, Bhawani (TK 8401) of toria, T 27 of taramira, GSL 1. **Mustard**: GM 1, GM 2, Laxmi, Narendra Rai, Pusajaikisan, Agmi (SEj2), jagannath (VSL 5), Pusa Bahar, Pusa Barani, PBR 91, PBR 27, Rajat, RCC 4, RH 819, RH 8113, RLM 1359, Samjukta, Asesh, Sarma, TM 2, TM 4. **Brown Sarson**: KBS3, KOS 1. **Yellow Sarson**: Pusa Gold,

Contd...

S. No.	Name of the crop	Varieties/Hybrids
		NOYS 921, Rajendra Sarson 1, Subinoy, YS 93.**Toria**: jawahar Toria 1, Panchali, RAUT 917, RH 68, TLC 1, TS 29.**Taramira:** RTM 314, TMC 1. Gobhi Sarson: GLS 2, PGSH 51, Sheetal.
26.	Rice/Paddy	Tulasi, Aditya, Prasanna, Rasi, Satya, Rudrama PNR 381, Birsadhan 101, Birsadhan 104, Birsadhan 1Q5, Birsadhan 201, Birsadhan 202, Heera, Kanchan, Kalinga III, Annada, GR 2, GR 5, GR 6, Himdhan, Nagardhan, VLK Dhan, VI. Dhan221, RP 2421, Himalaya 741, Himalaya 2216, Amrut, Mukti, lET 7564, Suyama, Modan, Onam, JR 75, R 28131-1, Tuljapur4, Ratnagiri 73-1-41, Teena, Kranti, Ngoba, Parijat, -Pathara, Kalyani II, Sattari, Neela, Rudra, Vanaprabha, Khandagiri, NiJagiri, Ghanteswari, Sneha, Patnai 23, Aswani. Renu, Saket 4, Narendradhan 18, Narendradhan 80, Narendradhan 97, Narendradhan 118, Khitish. Kiron, Bhupen. Swarndhan, Phalguna, Mandya Vijaya, Nandi, Swarna, Sambamahsuri, Pinakini, Krishnaveni, Thikkana, Chaitanya, Pothans, Or Jgallu, Sri Nanga, Sagarasamba, Simhapuri,Laksmi, Salivahana Bahadur, Kushal, Ranjit, Manoharsali, Mahsuri, Moniram Katekijoha, Rangili, Bhogali, Pankaj, Savitri, Sita, Jayashree, Janaki, Radha, CR 1002, Rajashree, Kanak, Vaidehi, Hemavathi, Nathravathi, lET 7191, Abhilash Intan, KPH 2, Neerja, Rashmi, Kayamkulam, Shyamalu, Sotril7, Kranti, Surekha, Mahamaya, Ratnagiri 2, SYE, 75, SYE-ER-1,Darna, Ratnairi 3, R 374-11, Jagannath" NEH Megha Rice I, NEH Megha Rice 2, Rajeswari, Seema, Parijat, CR1014, Jajati, Urbashi, Samalei, Pratap, Saradhi, Gauri, Daya, Mahalakshmi, Lakhmi, Savtri, Madhukar, Jauyalakshmi Suresh, Dinesh, Biraj, Bipsa, IR 42. Amulya, Nalini, Jogen sabita. Biraj, Jaladhi I, Suha, Jaladhi 2, Vaidehi, Jalmagna, Rambha, Utkal Prabha, Manika, Mahalaxmi, Kanchan, Panidhan, FR 13A, Jallahri, Jalnidhi, Jalapriya, Jitendra,Madhukar, Chakia 59, Natina, Mandira, Mahalakshmi. Purendu, Rasi, Vikas, Tella Hamsa, Prabhat, Abhaya, Rajendra Divya, Sasyashree, Bhaddrakali, Suraksha, IR 64, Ajaya, Vikramarya, Vibhava, Satua, Saleem, Surekha, Kavya, Erramalleu, Pusa 2-21, Pusa 44, Gautam, CR 1002, Archna, Rajendradhan 201, Rajendradhan 202, Jaya, Vikram, Karjat 2, Karjat 3, Ratnagiri 3, Sugandha, Ambika, Jaya, Narmada, Ratna, IR 20, IR 36, HR4, Gaur 10, GR 102, GR 103, HKR 102, Haryana Basmati, Tarogi Basmati, Basmati 370, Pusa 44, Puss Basmati 1, Kasturi PR 103, HKR 126, RP732, Himalaya 732, Ranbir Basmati, Jhelum, K78-13, SKAU 23, SKAU 27, Akash, Avinash, Kama, Mahaveer, Prakash, Vikram, Vidhava, Sonasali, IR 30864, Red Annapurna, Mandya, Vani, Triveni, Jyothi, Pavizkhani, Athira, Kartika, Makom, Sabari, Jayathi, Kanchana, Swamaprabha, Aishwarya, Aruna, Remya, Konakam, Mata Triveni,

Contd...

S. No.	Name of the crop	Varieties/Hybrids
		Kairali, Ruchi, Mahamaya, Patel 85 Madhuri Pawana, Sugandha, Punshi, Maniphouibl 1, Manephouibi 2, I R8, PR 106, PR 109, PR 108, Basmati 385, PR 111, Bharatidasan,Jawahar, Aravinder, Punithavathi, Savitri, Puduvai Ponni, BK 79, BK 190, Mahisugandha, IR 50, TKM9,. P~K1, ADT36, ADT37 , IR 20, Co 41, Co 43, Co 44, TPS1, MDU 2,MDU 3, ADT 39, ADT 42, JJ92, ASD17, White Ponni, Ponni, Paiyur 1, ADT 40, Co 42, Co 45, TRC BoroDhan, Saket 4, Sarjoo 52, Narendradhan 359 Govind, Madhukar, Jayalakshmi, Suresh, Dinesh, Biraj, Bipsa, IR 42, Givind, Pantdhan 4, Pandhan 6, Pandhan 10, Pandhan 12, Manhar, Narendradhan 2, Kunti,Lakhmi, Munal. CST 71, Lunishree, CSR 10, Panvel 1, Panvel 2, Vytilla, Vytilla 3, Vytilla 4, CSR 5, Vikas, CSR 5, Heera, Luit, Kalinga III, Pantdhan 1, Majhera 3, VL Dhan 39, VL Dhan 163, VL Dhan 206, VL Dhan 221, Ratnagiri 24, Jaya, Phule maval, Ambemohar 157, Endryani, Kundalika, Basumati 370, Karjat hybrid rice 1, Indrayani, Ambica, Terna, Parbhavati, Suganda, Parag, Parbhani Aviskar, Ratnagiri 73, Karajat 184, 1, 2, 3, 4, 5, 6, 14, 7, Ratnagiri 24, 711, 1, 68, 3, 2, Fondaghat 1, Shyadri 1, 2, 3, Palghar 1, 2, Panvel 1, 2, 3, Sakoli 6, 8, 7, PKV Ganesh, Shindewadi 1, 2, 4, 5, 75 and 2001.
27.	Safflower	Manjira, DSH 129, MKH 11, NARI 6, Sagara Muthyulu (APRR 3) (resistant to rust) A 1, APRR 3, DSH 129, MKH 11, NARI 6 Bhima, A 1, DSH 129, MKH 11, NARI6, Bhima, DSH 12, MKH 11, Nira (NRS 209), K 1, CO 1, HUS 305, T 65, EC 68414, SS 56, BSS 11, APSH 11, Modern, KBSH 1, LSFH 35, LSF 8, LS 11, BSH1.
28.	Sesame	Madhavi, Gauri, ST 168, RS 1, RT 1, OMT 11-6-3, B67, Kanke white, Krishna, Gujarat Til 1, Mrug 1, Purva 1, Haryana Til1, RT46, E8, TMV3, Co 1, Soma, Surya, Kayamkulam 1, Thilothama, Phule Til1, No.8, Tapi, TC25, JLT 26, N32, JT7, OMT 11-6-3 , TKG 21, Vinayak, Kanak, Kalika, OMT 11-6-3 (Uma), Pratap (C 50), TC25, T13, T46, TMV3, TMV4, TMV5, TMV6, Co. 1, SVPR 1 (TSS 6), T4, T12, RT46, B67, Punjab Till, Imp. Sel. 5 (Rama).
29.	Sorghum	CSH 1, CSH 6, CSH 14, CSH 9, CSH 10, CSH 11, CSH 13, CSH 16, SPH 388, AKSH 37, CSH 5, CSV 10, CSV 11, CSV 13, CSV 15, PVK 400, PVK 801, SPV462, S8 1066, DSV 1, GJ 36, GJ 37, GJ39, SPV 235, JJ 741, CSH 17, CSH 18, CSH 23, SPV 462, SPV 475, SPV 946, SSV 84, RSSV 1, PKV 400, CSV 15, PKV 801, PKV 809, SPV 655, SPV 1411, SPV 1595, SPV 669, SPV 839, Sel. 3, SPV 1359, RSLG 262, RSV 1546 (Phule Chitra), RSSGV 3 (Hurda sorghum), RSV 458 (Phule Anuradha), RSV 423 (Phule Vasudha), CSV 1626, RPOSV 3 (Pop sorghum), RSV 1006 (Phule Revati).

Contd...

S. No.	Name of the crop	Varieties/Hybrids
30.	Soybean	Portage, Flambeau, Merit, Grant, Traverse, Chippewa, Hark, Harosoy/63, Amsoy, Wayne, Adelphia, Clark-63, Kent, Delmar, Hill, York, Hood, Lee, Pickett, Davis, Bragg, Semmes, Hardee, Hampton, MACS 58, MACS 124, JS 335, PK 1029, MACS 450, DS 228 (Phule Kalyani), Parbhani sona, Jawahar, Samrudhi.
31.	Sugarcane	Co Pant 90223, B0120, Co B in 9605, Co 86249, Co 8371, Co M 88121, Co Jn 86141, Co 86032, Co 8014, Co 1148, Co 1158, Co 7717, B091, B099, CoJ 64, Co S687, Co S 767, C0419, Co 740, Co 975, Co 62175, Co 6304, Co 6007, Co 7219, CoC 671, Co 8021, Co 8013.
32.	Sunflower	BSH 1, APSH 11, LDMRSH 1, LDMRSH 3, KBSH 1, PSFH 67, PKVSH 27, DSH 1, EC 68414, EC 68415, Morden, Suraya, CO 1, BSH 1, KBSH 1, C02, SS56, TNAUSUF 7, GAUSUF15, PKVSF 9, APSH 11, LDMRSH 1, LDMRSH 3, PSFH 67, PKVSH 27, Bhima, Girna, Sharda, JLSF 414, S 4, DSH 129, MKH 11, Nari NH 1, Nari 6.
33.	Tobacco	FCVTobacco *Traditional deep black soils of AP:* Chatam, Dekrest. Kanakprabha, Dhanadayi, Jagasri. *Light soil area of Andhra and Karnataka:* Special FCV. CTRI Special. Godavari Special (resistance to TMV), Swarna (resistant to powdery mildew), McNair 12 (tolerant to black shank), Jayasri (MR) (resistant to TMV), K 326 and Trupthi (NLS 4) (tolerant to black shank and nematodes. Bidi Tobacco *Gujarat:* Keliu 20 m Anand 3, Anand 23 (tolerant to leaf-borer), GT 5 and GTH 2 (tolerant to root-knot nematode), G T 7 (tolerant to drought) *Karnawka and Maharashtra and Andhra Pradesh:* Anand 2. Anand 119 *Karnawka and Maharashtra:* NPN 190, PK 5 Chewing Tobacco *North Bengal:* Chama (clay soils); Podali (snady soils) *Bihar:* DP 401, Gandak Bahar, Sona, Prabha, Pusa Tobacco 76. Vaishali Special *Tamil Nadu:* Vairam (pit-cured areas), Thangam and Maragadham (smokecured areas), Bhagyalakshmi and Meenakshi (sun-cured tobacco) Rustica Tobacco *West Bengal (Motihari):* DO 437, Sonar Motihari *Gujarat:* GC 1, GCT 2 Natu Tobacco. *Andhra Pradesh:* Prabhtit (resistant to TMV); Vishwanath, Natu special Cheroot Tobacco *Andhra Pradesh (river ~ island of East Godavari):* DR 1, Lanka special *Tamil Nadu:* Bhavani Spedal (Coimbatore), Sendarapatty Special (Salem) Cigar - Wrapper Tobacco *Tamil Nadu:* Krishna Burley Tobacco *Andhra Pradesh (light soil agency areas of East Godavari, Vtshakhapatnam and Vtjanagaram district):* Banket Al (resistant to TMV).
34.	Wheat	PBW 343, HD 2687, WH 542, UP 2336, Raj 3077, CPAN 3004, PDW215 (d) PDW 233(d), WH 896(d). PBW 373, UP 2338, PBW 226, Raj 3765, Raj 307, UP

Contd...

S. No.	Name of the crop	Varieties/Hybrids
		2425, HD 2285, Raj 3777, Raj 3765, HD 2402 PBW 175', PBW 299, PBW 65, PBW 396, WH 533 (in Haryana), C 306, HD 2733, K 8804 (K 88), HD 2402, HP 1731, HP 1761, K 9107 (Deva), PBW 443, NW 1012, HUW 468, HD 2643, HP 1633 (Sonali), DL 784-3, HP 1209, HUW 234, HP 1744, NW 1014 HD 2285, Raj 3777, Raj 3765, HD 2402 K 8027, C 306, HDR 77, K 8962 (Indra), K 9465 KRL 1-4, Raj 3077, KRL 19, Job 666, DL 803-3, GW190, HI 1077, HI 8498 (d), GW 273, HD 2236, Raj 1555 (d), HI 8381 (d), GW 173, Swati, J 405, DL 788-2, HD 2285, Raj 3777, Raj 3765, HD 2402 C 306, Sujata, A 9-30-1 (d), HW 2004, JSW 17, HD 4672 (d), KRL 1-4, Raj 3077, KRL 19, Job 666, MACS 2496, DWR 162, DWR 1006 (d), HD 2380, MACS 2846, AKW 1071 (in Maharashtra), HD 2501, DWR 195, HI 977, NIAW 34 NI5439, N 59 (d), MACS 1967 (d), Bijaga yellow (d), NIDW 15 (d), KRL 1-4, Raj 3077, KRL 19, Job 666 NP 200, DDK1001, DDK 1009, MACS 3125, NIAW 301, NIAW 295, NIDW 15, N 5439, N 8223, Trimbak, Panchwati, Godawari.

REFERENCES

1. Acquaah, T. (2002). *Principles of crop production*. Prentice-Hall of India, New Delhi.

2. Panda, S.C. (2005). *Agronomy*. Agrobios (India) Jodhpur. p. 833.

3. Panda, S.C. (2004). *Cropping and Farming Systems*. Agrobios (India) Jodhpur. Pp. 833.

4. Prasad Rajendra (2004). *Textbook of Field Crops Production*. ICAR, New Delhi. Pp. 821.

5. Reddy, S.R. (2006). *Agronomy of Field Crops*. Kalyani Publishers, New Delhi. Pp. 698.

6. Reddy, S.R. (2006). *Principles of Agronomy*. Kalyani Publishers, New Delhi. Pp. 611.

7. Balasubramaniyan, P. and Palaniappan, S.P. (2001). *Principles and Practices of Agronomy*. Agrobious (India), Ludhiana.

8. Beets, W.C. (1982). *Multiple cropping and tropical farming systems*. Westview Press, Boulder, Colorado.

9. Chakraverty, A. (1995). *Post harvest technology of cereals, pulses and oilseeds*. Oxford and IBH Publishing Co., Pvt. Ltd. New Delhi.

10. Chatterjee, B.N. and Maiti, S. (1984). *Cropping systems-theory and practice*. Oxford and IBH Publishing Co., Pvt. Ltd. New Delhi.

11. Norman, D.W. and Collinson, M.P. (1985). *Farming systems Research: Theory and Practice*. ICIAR Proc. No. 11: 16-30.

12. Singh, S.S. (2007). *Crop management under irrigated and trained conditions*. Kalyani Publisher. New Delhi.

13. Somani, I.I. (1992). *Dictionary of weed science*. Agrotech Publishing Academy. Udaipur, India.

14. Gangopadhyay, A. 2007. *Crop Production Systems Management*. Gene-Tech Books, New Delhi.

15. Singh, R.P. 2007. *Sustainable Development of dryland Agriculture in India*. Scientific Publishers (India) Jodhpur.

16. Bhattacharyya, P. and Purohit, S.S. 2008. *Organic Farming*. Biocontrol and Biopesticide Technology. AgroBios (India).